The All-Day Energy Diet

Hay House Titles of Related Interest

YOU CAN HEAL YOUR LIFE, the movie, starring Louise Hay & Friends
(available as a 1-DVD program and an expanded 2-DVD set)
Watch the trailer at: www.LouiseHayMovie.com

THE SHIFT, the movie, starring Dr. Wayne W. Dyer
(available as a 1-DVD program and an expanded 2-DVD set)
Watch the trailer at: www.DyerMovie.com

III

ARE YOU TIRED AND WIRED?: Your Proven 30-day Program for Overcoming Adrenal Fatigue and Feeling Fantastic Again, by Marcelle Pick

CRAZY SEXY KITCHEN: 150 Plant-Empowered Recipes to Ignite a Mouthwatering Revolution, by Kris Carr

THE FATIGUE SOLUTION: Increase Your Energy in Eight Easy Steps, by Eva Cwynar MD

HEALING WITH RAW FOODS: Your Guide to Unlocking Vibrant Health Through Living Cuisine, by Jenny Ross

THE POWER OF SELF-HEALING: Unlock Your Natural Healing Potential in 21 Days!, by Dr Fabrizio Mancini

SLIMMING MEALS THAT HEAL: Lose Weight Without Dieting, Using Anti-inflammatory Superfoods, by Julie Daniluk RHN

THE TAPPING SOLUTION: A Revolutionary System for Stress-Free Living, by Nick Ortner

All of the above are available at your local bookstore,
or may be ordered by contacting Hay House (see next page).

III

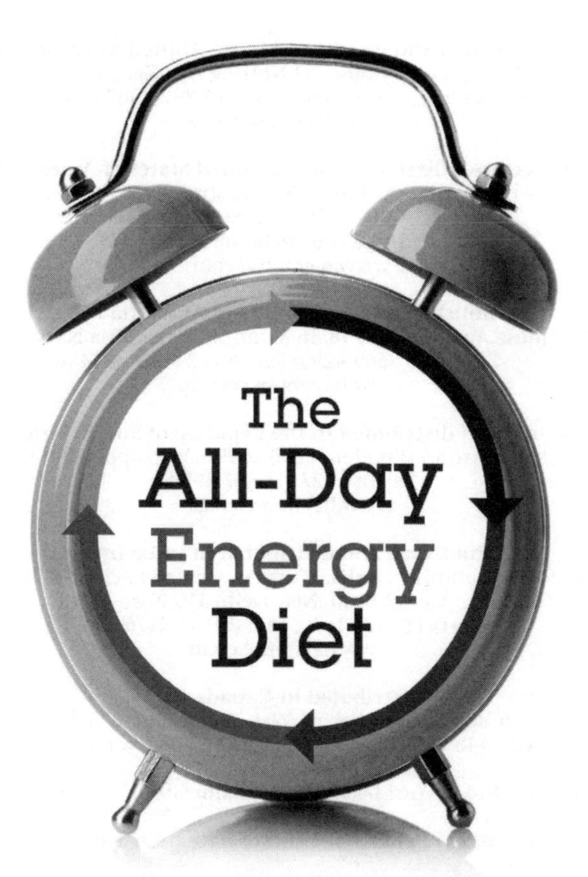

Double Your Energy in 7 Days

YURI ELKAIM

HAY HOUSE

Carlsbad, California • New York City • London • Sydney
Johannesburg • Vancouver • Hong Kong • New Delhi

First published and distributed in the United Kingdom by:
Hay House UK Ltd, Astley House, 33 Notting Hill Gate, London W11 3JQ
Tel: +44 (0)20 3675 2450; Fax: +44 (0)20 3675 2451
www.hayhouse.co.uk

Published and distributed in the United States of America by:
Hay House Inc., PO Box 5100, Carlsbad, CA 92018-5100
Tel: (1) 760 431 7695 or (800) 654 5126
Fax: (1) 760 431 6948 or (800) 650 5115
www.hayhouse.com

Published and distributed in Australia by:
Hay House Australia Ltd, 18/36 Ralph St, Alexandria NSW 2015
Tel: (61) 2 9669 4299; Fax: (61) 2 9669 4144
www.hayhouse.com.au

Published and distributed in the Republic of South Africa by:
Hay House SA (Pty) Ltd, PO Box 990, Witkoppen 2068
Tel/Fax: (27) 11 467 8904
www.hayhouse.co.za

Published and distributed in India by:
Hay House Publishers India, Muskaan Complex, Plot No.3, B-2,
Vasant Kunj, New Delhi 110 070
Tel: (91) 11 4176 1620; Fax: (91) 11 4176 1630
www.hayhouse.co.in

Distributed in Canada by:
Raincoast Books, 2440 Viking Way, Richmond, B.C. V6V 1N2
Tel: (1) 604 448 7100; Fax: (1) 604 270 7161; www.raincoast.com

Interior design: Nick C. Welch

A catalogue record for this book is available from the British Library.

ISBN: 978-1-78180-393-6

Printed and bound in Great Britain by TJ International, Padstow, Cornwall.

Dedicated to the loves of my life:
Amy, Oscar, Luca, and Arlo. You make having
more energy necessary and immensely rewarding.

CONTENTS

Preface ix
Introduction: From Soccer to Bed to No Hair on My Head xiii
The 7 Commandments of Energy xxi

PART I: The Extreme Fatigue Epidemic

Chapter 1: Why Are We So Tired? 3
Chapter 2: 3 "Sneaky" Foods That Rob Your Energy 19
Chapter 3: The Untold Secret Inner Workings of Your Body 31
Chapter 4: The Dirty Daily Habit That Makes You Sluggish . . . and How to Fix It 41

PART II: The Energy Resurrection

Chapter 5: What to Eat for All-Day Energy 57
Chapter 6: The 7-Day Energy Reset 105
Chapter 7: Safe Supplements and Simple Superfoods That Give You Lasting Energy 123
Chapter 8: Digestion Damage Control (and How to Avoid the Dreaded Food Coma) 139
Chapter 9: Energizing Exercise That Burns Fat and Doesn't Crush Your Body 153
Chapter 10: 9 Ways to Ease Stress and Balance Your Hormones 167
Chapter 11: Finding Your Passion and Purpose to Feel Awesome 189

Endnotes 197
Recommended Resources 207
About the Author 209

PREFACE

I was sick and tired, and had lost nearly half the hair on my head. And I was only 17.

You'd think that as an active and fit teenager, I would have been in my prime—full of energy, healthy, and ready to take on the world. At the time, I thought I could do just that, but it was going be tough, considering that I was spending 12 to 14 hours a day in bed sleeping. Even with those precious hours of rest, I could barely get by.

What happened to me is by no means rare. It's happening to millions of people right this very moment. Quite honestly, I'm glad that I went through that ordeal, because it unleashed within me a compelling desire to help others find greater health and vitality in their lives. I want you to be one of them.

My name is Yuri Elkaim, and I'm on a mission to empower at least ten million people to experience all-day energy and amazing health in the next few years. In just a moment, I'll tell you more about my health challenges, my life-sucking low energy levels, how I overcame them, and how you can do the same in just a few short days. First, however, I want to share what I believe to be true when it comes to health. These beliefs stem from having helped hundreds of thousands of people achieve amazing health over the past 15 years.

Good health is the path to a life of freedom and joy. Furthermore, every single one of us has only scratched the surface of our true health potential. There are many reasons for this, and as I see it, all of them can be summed up with one explanation: we don't understand or appreciate the importance of energy.

Above all else, I value energy. Although many of us take it for granted, feeling energetic is the ultimate sign of great health. It's the essence of truly feeling alive. When you're firing on all cylinders, your mind is sharp and you're physically able to tackle any task that comes your way. Beyond that, the challenges you face at work and in your relationships don't hold you back; rather, they help you reach new levels of accomplishment and happiness.

Without energy, however, your days drag on, with no zest and little joy, and fizzle into an exhausting finish. That's no way to live, but unfortunately, that's exactly the kind of existence most of us settle for every single day. It shouldn't be like this. Each day is a gift, and we need to live each moment to the fullest—giving the best of ourselves and doing what matters most in our lives.

I am committed to helping us overcome this grim state of affairs. We all deserve healthier, happier lives, but for this to happen, we have to change our approach to nearly everything we do: how we eat has to shift, how we exercise has to evolve, and we have to establish reliable ways to deal with our stress. Without a complete overhaul of these areas of our lives, we cannot hope to truly feel as amazing as we're meant to.

While that task may sound daunting, you'll find a simple blueprint to this "energy makeover" in these very pages. Before you can get started, though, we have to acknowledge that almost everything that we've been told or sold about what it takes to live these happier, healthier lives is flat-out wrong. The North American health-care industry is often more concerned with researching cures than providing us with actionable ways to prevent the diseases and conditions so many of us fall victim to. Although I'm a geek at heart, the findings of modern health science mean very little to me unless they truly make you feel better and improve your health in the process. After all, you can pretty much find scientific literature to support any point of view nowadays.

I've also become convinced through research and personal experience that much of what we eat and drink directly contributes to our poor health and nearly nonexistent energy levels. Even those products that are marketed as healthy are anything but, pumped full of preservatives and chemicals that hurt us in the long run. On a very basic level, the foods we eat should make us feel better than before we ate them. Falling asleep after a meal shouldn't be considered the "norm," nor should feeling exhausted all the time. Sinking into lethargy after every meal is a ticket to a life of tired desperation.

As a former athlete and personal trainer, I believe that getting in shape shouldn't be something we pursue simply so we can live up to societal standards of how we "should" look. Fitness isn't about looking good in our underwear. Is it any wonder so many of us have such a begrudging attitude toward working out? We should get excited about how good it makes us *feel*. For this to happen, how we exercise has to change so that working out leaves us feeling invigorated and energized, not drained. It should make us stronger, not more prone to injury.

Finally, I am convinced that the *way* we work and provide for our families is corrosive, merely giving us the money we need to pay our bills and little more. It saps us of our natural energy.

Your work should excite you. After all, you will probably spend the majority of your adult life at your job; shouldn't it be something that inspires you and leaves you feeling fulfilled at the end of the day?

■ ■

Throughout our time together in this book, I'm going to share some truly revolutionary (and perhaps contrarian) ideas, insights, and strategies that will help you break the damaging patterns and enjoy the all-day energy you're seeking. As I often tell my clients, "How you heal anything is how you heal everything." Thus, the path to enjoying more energy will also allow you to lose weight almost effortlessly and improve most aspects of your health. You can do it all, and it's much easier than you think.

This is the path to fixing and improving so many areas of your life that just don't seem to be working. This is how you mend yourself.

And just to be clear—this has nothing to do with energy shots or other dangerous gimmicks. This is not a fly-by-night health fad promising too-good-to-be-true results. These are fundamental health strategies that, when applied, work 99 percent of the time for 99 percent of people. No matter what type of diet you follow, you can integrate these strategies into your life. As long as you're a human being from planet Earth, applying any of the ideas you discover in this book will change your life for the better. I guarantee it. I've seen it more times than I could possibly mention or you could possibly imagine.

Sound good?

Great.

■ ■ ■ ■

INTRODUCTION

From Soccer to Bed to No Hair on My Head

Before we begin, I need to lay all of my cards on the table. All I believe and have come to know about health emerged from my painful and embarrassing past. Simply typing these words makes me feel vulnerable and brings back the shame I once felt. In fact, it's so bad that I almost removed my personal story from this book. Nonetheless, I'm glad to be sharing it with you because there are some valuable lessons to be learned.

I wasn't always a best-selling health-and-nutrition author or a highly sought-after holistic nutritionist. I certainly wasn't always a guy with a balanced diet.

In fact, for around two decades, there was always something wrong with me. I was dead tired and suffered from an avalanche of health problems. I endured all of this even though I was an athletic and supposedly fit young man. I tried to ignore it, but eventually there came a day when I decided enough was enough.

It was late one Wednesday night in the middle of March, a few days before my 17th birthday. I had just returned home from my regular soccer practice, but I was even more of a mess than usual. It was quite warm for that time of year, and I remember how sticky I felt as I peeled off my sweaty training attire. I was a goalie, so my padded shorts and knee-high socks were caked with dirt. I'd even picked up a few skid marks along my left thigh from all the diving around in the mud. As I said, I was a mess, but that was nothing compared to what was to come.

I felt ready to collapse as I dragged my body up the never-ending staircase of my family's three-story home and made my way into the bathroom to get cleaned up. I always felt tired and gross after practice, but this was something different: my breathing was labored, and my hands were throbbing as they never had before. You see, having to wear sweaty goalie gloves a few times a week used to make my lifelong eczema flare up like molten lava oozing out of a volcano. It was so horrible that even on the warmest spring days, I would wear winter gloves on the school bus to hide my reptilian bare hands from my classmates. Alone in the bathroom, I stared at my red, scaly hands under the vanity lights and felt more disgusted with myself than I had ever been before.

As if my inflamed hands weren't bothersome enough, I could barely breathe, so I decided to take a hit from my steroid puffer to settle my asthma. It always acted up during spring training, but tonight was particularly rough. In fact, it reminded me of my epic asthma bouts as a young child, the ones that would find me strapped up to a vaporizer mask injected with steroids just to keep my airways open. Back then, playing sports with my friends was close to impossible, and I wondered now if things were getting that bad again. It was even making me question my future in soccer, the true love of my life. I so desperately wanted to last an entire training session or 90-minute game without keeping my puffer right beside the goalpost "just in case." It was beginning to seem impossible.

As I stared at my reflection in the bathroom mirror, I felt like I belonged in a horror movie. My life was slowly coming apart because of my health issues. On top of my skin and breathing problems, I barely had enough energy to make it through the day, even when I wasn't playing soccer. I'd wake up each morning feeling as if I'd just downed an entire bottle of tequila the night before. I was sick and tired of feeling like I was 30 years older than I really was. I was 17, damn it!

Once I caught my breath a little bit, I stepped into the shower to wash off the sweat, grime, and self-loathing. I yanked the shampoo from the basket hanging from the showerhead, squirted the gel into my hands, and began lathering up my hair. As the warm water coursed through my hair, a feeling of rejuvenation and hope washed over me. I closed my eyes and let out a deep sigh.

My relief didn't last very long.

When I opened my eyes, I looked at my hands and saw them covered in thick clumps of my brown hair. It was everywhere! What was going on?

Shocked and confused, I immediately jumped out of the shower and toweled off. I noticed even more hair sticking to the fibers of my wet towel. In a panicked frenzy, I rushed over to the mirror to investigate more closely.

I just couldn't understand what I was seeing. I had a bald patch over my left ear the size of a quarter. I hadn't noticed it until now since my longer hair normally covered that spot on my head. I guess it must have just started to fall out. It didn't make sense, and I was scared. No one in my family had had any hair-loss problems, and certainly not at 17 years old. What was happening to me?

The next morning, I woke up to a pillow covered in hair. Just a few weeks later, all of the hair on my head had fallen out. My eyebrows and eyelashes had vanished as well.

Needless to say, I felt like a freak. At school I endured the bewildered and disgusted stares of my classmates and even my soccer teammates. Maybe they thought what I had was contagious; I don't know. I wanted to shut myself in my locker and never come out.

I honestly don't know how I would have made it through that difficult time without my family and close friends. I will always be indebted to them. As difficult as my health problems had been before, this was my absolute rock bottom. It was pretty freaking low.

But you know what?

As bad as all that was, I'm truly grateful that I went through it. If I hadn't, I wouldn't be in a position to share my life-changing wisdom with you. That horrifying night after soccer practice and the ensuing events gave me a much-needed wake-up call. It was that ordeal that inspired me to make some serious changes in my life.

I made a promise to myself that I would get to the root, so to speak, of my unexplained hair loss, as well as my lifelong asthma and embarrassing eczema, and finally put an end to feeling tired all the time. I was going to figure it all out or die trying.

My solution? I decided to consult with every kind of doctor I could find to figure out how I could heal myself. First up was my immunologist, who told me that I had an autoimmune condition called alopecia. It should have been illuminating, yet he had no answers beyond that diagnosis. How could that be?

Over the next seven painstaking years, I discovered a pattern as I worked my way through every corner of the medical community—not a single doctor had any idea of what to do other than to inject cortisone directly into my scalp. Not one doctor or "specialist" ever speculated as to what might have caused my health problems to begin with or what could be done to solve them. They were more interested in getting me on experimental pills and steroid creams or injecting my scalp with who knows what. There were no real answers. My frustration knew no limit.

And so, by the time I was 24 years old, I made my second life-changing decision.

Realizing that neither the medical system nor any doctor within it was going to help me back to better health, I vowed to read every book and crawl under every rock, if necessary, to learn about health so I could finally heal myself. If the broken medical system couldn't do it, then I would figure it out on my own.

My first step was to enroll at the University of Toronto and attain a degree in physical education and health/kinesiology. This was a good start, but I still hadn't fully grasped how to be healthy. That would begin to change shortly afterward. But before that change I somehow found the energy to play soccer for the varsity team for four years, while working as a personal trainer and taking on a full course load. Yes, I guess I'm a little type A.

As much as I loved school and training clients, my ultimate goal was still to play pro soccer in Europe. Thankfully, that sort of happened thanks to the great relationships my agent had with various clubs in France.

So, after college, my agent sent some of my video footage along with a glowing recommendation over to one of the top clubs in France. They were impressed and invited me to come play with them on a "probationary" contract. I was moving to France to pursue a career as a professional soccer player!

But as a goalkeeper (and third string) I spent most of my time playing with the reserves for the club. It takes years for most goalkeepers to crack into the first team. Nonetheless, I was in the thick of it.

It was exciting, but I soon found myself faced with my asthma again. As if that wasn't enough, I quickly became disillusioned by pro soccer, mainly because of how little control I had over my career path and because I was overly critical of my own performances, which meant I spent a lot of time feeling unhappy. I nursed my sorrows with French bread, croissants, and coffee, and my sluggishness soon returned.

Deep down inside, I had a feeling that my health issues might have been related to my diet, which was very unbalanced, to say the least. For much of my life, I'd been strung out on grains, lots of sugar, endless dairy, and every type of processed food you can think of. Rarely would you catch me eating something as light, nourishing, and refreshing as a piece of fruit. As for veggies—what were veggies?

After spending just one season in France, I had an epiphany that this wasn't the life I wanted to live. I felt compelled to do something greater (like writing this book, for instance) and actually make the world a better and healthier place. And so, I decided to hang up my boots and retire from pursuing a career in pro soccer. *Greener pastures, here I come,* I thought.

■ ■

Sadly, my poor eating habits stuck with me when I left France and returned to Canada. Soccer was now a personal hobby, not a vocation, and I still found myself haunted by the health problems that had plagued me for so long. *Was it really my diet?* I couldn't escape this nagging question.

Then one day, I saw an advertisement for an open house at a holistic nutrition school in Toronto. I knew I had to attend.

When I showed up at the school and started digging into all it had to share, something inside of me clicked. In fact, within just ten minutes of being there, I started to discover the answers I had been desperately trying to find for so long. I was floored by what I was learning!

I was hooked, so I enrolled immediately. It got even better when I asked one of my soon-to-be-professors—an incredible naturopathic doctor—whether or not my autoimmune condition, constant fatigue, eczema, and asthma could have been caused by my poor diet.

Her answer: "Absolutely! We see it every day."

I couldn't believe it! Somehow, a solution to my problems existed, and I was about to embark upon an amazing (and surprisingly quick) journey to a full recovery. In four years at one of the top universities in the world, I had never been taught this information. It struck me that if *I* didn't know this stuff, there must be billions of others around the world still left in the dark. I had found my calling.

I eagerly applied what I was learning (much of it you'll discover in this book) and noticed my hair starting to grow again. I was feeling more energetic. And my skin problems and asthma had vanished. I was super-excited.

Not feeling drugged upon waking, and soaring through my days with lasting energy, was a huge benefit. It allowed me to spend less time sleeping and more time getting stuff done. It's allowed me to be more productive, help more people, spend more time with my kids, and still pursue my other passions and hobbies.

Roughly two months after starting my studies in holistic nutrition, most of my hair had returned, although since that time my hair has come and gone a few times. That just seems to be the way alopecia works. Nonetheless, figuring out how to outwit my breathing, skin, and energy problems was a huge victory for me.

With a spring in my step and a world to empower, I took the information I learned at this school and combined it with my athletic and fitness background. Armed with all of this, I created an approach to healthy eating, exercise, and life that would eventually make me one of the top health-and-fitness experts in our community.

Everything I learned along this incredible journey is what I have to share with you.

I hope this shows you that despite my success today and the fact that people now see me as the go-to energy-nutrition expert, I'm no different from you. I'm just a guy who went through hell and stumbled upon some very real solutions in the middle of it all. These solutions are what I'm offering you.

This is a plan that allows people like you and me to experience sustained energy all day long and a healthy body that we can feel proud of.

This is the same plan that has allowed me and thousands of my clients to effortlessly spring out of bed in the morning and surge through life with nonstop energy and focus.

With the plan in this book, you can finally see the results that have eluded you for so long. Specifically, we'll be diving into seven areas that must be addressed in order for you to reach your energy and health breakthrough. They are:

1. The pH of your blood
2. The health of your digestive system
3. The state of your adrenals
4. Specific foods to eat
5. Helpful supplements to take
6. Simple stress-management techniques
7. The proper way to exercise to avoid killing yourself any further

This is the plan that will change your life.

It certainly has for me, considering how much stuff I've got going on. For instance, by my 30th birthday a few years ago, my first son, Oscar, had arrived. Less than three years later, two new additions, Luca and Arlo, had graced us with their presence. With three little boys and a compelling health mission, I need all the energy I can get. How about you?

If you're like me and you've got kids and a ton of other commitments (work, friends, hobbies, and so forth), then having more energy is the key to unlocking an unbelievable life. Imagine getting out of bed in the morning and actually feeling awake. Imagine getting through your workday without ever feeling a slump. Imagine coming home in the evening and still having plenty of energy to play with your kids. How much of a difference would that make in their lives? How much of a difference would it make in yours?

See yourself at work or in school with laserlike focus. You're in the zone, and you're far more productive. It's almost as if your body is on autopilot, and those days of caffeine crashes and sugar cravings are behind you. Finally, you're in full control. You feel great. You look great. And the best part is that you've accomplished this in just a few days' time, by applying a handful of simple health strategies.

I don't know about you, but I'm super-excited for you. I'm so pumped that I can barely stay seated! You are about to experience utter and absolute transformation. You'll be able to tell the world about how you've awakened the true you and no longer have to settle for mediocrity. You deserve to look and feel amazing. You really do.

Let's get started. . . .

■ ■ ■ ■

THE 7 COMMANDMENTS OF ENERGY

Over the years, I've made some interesting observations about what makes us feel energized. Some of them are scientifically proven, while others remain theoretical. Nonetheless, these observations have become my energy principles, and they provide a solid foundation for understanding what you're about to learn and apply throughout this book. Think of these as the 7 Commandments of Energy. As you go through *The All-Day Energy Diet,* these principles will make more sense to you. For now, I just want to lay them out for you.

Principle #1

Since your blood is the river that carries oxygen to your cells for energy production, any food or process that compromises the health of your blood or ruins your oxygen-carrying red blood cells will deplete your energy.

Principle #2

Because digestion is such an energy-dependent function, any food or process that imposes more demand on, or compromises, your digestion will drain your energy. Likewise, any food that provides more energy (not calories) than it requires to digest will make you feel more alive.

Principle #3

Since digestion consumes so much energy, eating *less,* paradoxically, can give you *more* energy (and has been shown to be one of the few ways to extend your life).

Principle #4

Your adrenal glands help you deal with stress. Excessive stress (attributable to food, exercise, lifestyle, or what have you) that is not managed properly and eventually becomes chronic will weaken your adrenal glands and impair your ability to cope, ultimately depleting your energy.

Principle #5

There seems to be a relationship between cognitive function and energy. A cluttered mind, depressed emotional state, and inability to focus tend to drain your energy. Thus, any activity that clears and focuses your mind can give you sustained energy.

Principle #6

Movement is life. Stagnation (rigor mortis) is death. A consistent lack of energy indicates that your body is diverting more of its resources away from movement, inward for internal healing and repair. Think of how lethargic you felt when you were sick or if you've ever been pregnant.

Principle #7

How you heal anything is how you heal everything. The things you do to improve your energy will also improve your overall health. Weight loss becomes easier, performance improves, and you lower your risk for disease.

■ ■

Once again, these are my observations. Whether or not you believe them doesn't matter, because once you see them in action, there will be little argument. I'm not the energy expert for nothing! As you can tell, this is not your run-of-the-mill diet book. It's loaded with both proven and contrarian principles and strategies that work.

The following chapters will delve deeper into each principle, providing "Aha" moments and equipping you with numerous actionable tips that have already transformed tens of thousands of my All-Day Energy Diet clients all around the world. Be sure to take action on at least one thing from each chapter. If you do, I can guarantee you will double your energy in about seven days. If you *don't,* then you will no doubt continue struggling with low energy.

Now, before we dive in, I have two simple challenges for you:

1. Read this book in its entirety. You don't have to do so in one sitting, but you don't want to miss a single chapter, because there are hidden gems in each one that could transform your life.

2. Keep an open mind. I'm going to challenge you and what you've been led to believe is true about nutrition and the human body. Keep your mind open to the possibilities that exist beyond the scientific literature and popular media. You'll see what I mean, especially in the chapter on food.

Please keep these commandments in mind as we go through this journey together. As we venture into Part I, "The Extreme Fatigue Epidemic," you'll quickly see how Principle #1 plays a major role in your day-to-day energy levels.

Without any further ado, let's get into it.

■ ■ ■ ■

Part I

THE EXTREME FATIGUE EPIDEMIC

Chapter One

WHY ARE WE SO TIRED?

Life is supposed to be challenging, but it's not meant to seem impossible. It surely feels that way sometimes, though, doesn't it?

If I were to ask you whether or not you'd like more energy, I'm certain the answer would be an emphatic, nearly desperate *yes!* Then again, you're reading this book, so that kind of gave it away. It's not surprising, as having more energy and feeling more alive every single day is one of the most common desires I've noticed among all people. Even those who don't say they crave more energy are probably fooling themselves. Some of them may say they want to lose weight, while others want to gain muscle. Some people say they want better-looking skin, and others want to be better performers in their sport or area of work. What's the one thing that makes any of these goals so much easier to achieve?

Energy.

Not only does being more energetic allow *you* to get so much more out of life, it also allows you to contribute so much more to your friends, family, and community. For me, the latter is a biggie. As nice as it was to overcome many of my personal health challenges several years ago, the most important improvement in my life was my increased energy levels. Instead of sleeping 10 to 12 hours a night, my body only required 6 to 8 hours of sleep, and I woke up most mornings ready to conquer the world! I was focused and productive all day long, which meant I was able to help more people and get my message out to the world. It's tough to do that when you're sidelined on the couch feeling exhausted.

I was in peak condition for quite a while, but it didn't last. With the arrival of my kids several years later and a more hectic lifestyle, I noticed my energy levels regressing. I wasn't waking up as easily as I had just a few months prior, and I just didn't have the lasting daylong energy I previously enjoyed. Exercise became a chore; I just couldn't push myself to the same level of intensity that I had before. What made it more confounding is that I was eating well (according to the principles you'll discover further along in this book), but something still wasn't quite right.

Once again, I set off on a quest to figure out what was going wrong inside my body. After much research, visits with my naturopathic doctor, and several lab tests, I discovered that my adrenal glands were shot. (I'll explain what this means in more detail a little later, but essentially, my body was no longer able to cope with stress very well.) Armed with this new knowledge, I was able to make some great progress in regaining my all-day energy and bringing my adrenals back to life.

The vital information I learned along the way can be found in the pages of this very book, and it can help *you* overcome the drain that you're facing. Don't you want to be able to jump out of bed in the morning and coast through your days with boundless energy? What if you could avoid that dreaded midafternoon slump every day, the one that leaves you completely spent after work, unable to play with your kids or go to the gym? Wouldn't you like to feel downright *awesome?*

You can, but first you have to understand why your energy is being sapped away.

■ ■

Energy is life. An energetic body is a healthy body. Conversely, a tired and lethargic body is groaning and wheezing for help. Without sufficient energy we jeopardize our health, accomplish a fraction of what we're truly capable of, and drag through our days like zombies grasping for quick fixes that offer temporary, if questionable, help.

What am I talking about? That would be caffeine, the number one drug in the world that's consumed by more than 100 million Americans each day in the form of coffee and almost as often in the form of energy drinks. According to a report by Agriculture and Agri-Food Canada in 2008, 1.5 *billion* cans of Red Bull were sold in the United States in 2004. The U.S. energy-drink market alone exploded by more than 240 percent from 2004 to 2009, and is predicted to at one point reach an astounding $19.7 billion.[1]

The irony is that hundreds of millions of people are still drained of energy. In fact, we are more tired than ever before. Do any of the coffee drinkers or energy-drink guzzlers you know—and you might be one yourself—truly seem pumped full of the energy they need to get through the day? We are overstressed and overworked, and in spite of being over-caffeinated, we are still struggling to get out of bed in the morning and take care of our to-do list. Obviously, what we're doing is not working.

The reason is simple—culprit beverages (namely, energy drinks, coffee, and soda) hit our bodies with a strong dose of caffeine, which wears down our adrenal glands and wreaks havoc on our blood-sugar levels. Out-of-whack hormones and dwindling energy are the results. Aside from the short-lived pick-me-up we might experience from these stimulants, the long-term consequences are truly scary. We'll investigate these consequences in more detail a little bit later.

Right now, we citizens of "Caffeination Nation" face an extreme fatigue epidemic. The number of those affected reaches into the hundreds of millions. This is a problem of cataclysmic proportions, and it is robbing everyday people (perhaps even you) of the happiness, productivity, libido, and fulfillment that they deserve. It has to stop.

It's for this reason that I've written *The All-Day Energy Diet.*

The food most of us eat doesn't help, either. Since mass food processing began in the 1950s, we have traded our health for convenience by relying on fast food and packaged, processed foods devoid of nutritional value. They may taste great, but they're no good for the human body. They're dead foods, and since we become what we eat, we *feel* dead and drained of life and energy. Instead of nourishing us, this commercial "food" often compromises our digestive and intestinal health, acidifies our blood, and demands more energy from our bodies' inherent processes to do so.

We are sicker, fatter, and more tired than ever before. Thankfully, the solution has nothing to do with eating more protein, cutting out carbs, or jumping on the latest Hollywood diet fad. The solution involves getting to the root of the problem, cleaning up our bodies' internal mess, and reconsidering what we put into our mouths. It may sound like a nearly impossible task, but it's actually quite simple when done correctly. It's this time-tested solution that I reveal in the chapters to come.

We'll soon explore the major players that are robbing you of the energy you deserve (for instance, "dead" foods, poor lifestyle factors, and the like), but in order to understand them, we need to first have a better grasp of how your body works.

How Your Body Works and Copes with Stress (in a Nutshell)

There are countless processes that take place in the amazing machine that is your body. Let's take a look at a few of the most important ones that directly impact your energy levels. In explaining these to you, I'm going to leave my lab coat at the door and discuss them in very common terms so that they actually make sense to you.

The first thing you need to understand is how your body handles stress. That function begins in your adrenal glands, two walnut-size glands that sit above your kidneys. Although they're very tiny, the adrenal glands are responsible for one of the most important human behaviors, the fight-or-flight response. Chances are you've heard of it before, although you might not quite be familiar with what it involves. Here's what happens:

Your brain perceives a threat. Within milliseconds, nerve impulses are fired from down your spinal cord and your sympathetic nervous system is activated. This results in dilated pupils, increased heart rate, opened airways, and other reactions that instantly prepare your body to fight or flee. Seconds after this initial neural response, hormonal and nerve impulses are sent from your brain to your adrenal glands. Their message? Secrete adrenal hormones—cortisol and epinephrine—to quickly break down stored fuel into blood sugar for immediate energy. This brain-to-adrenal-gland communication is known as the *hypothalamus-pituitary-adrenal axis* (HPA axis).

Without this incredible mechanism, humanity probably wouldn't even exist anymore. Why? Well, imagine for a moment that it's 10,000 B.C. and you and your tribespeople are roaming the plains of the midwestern United States. Let's assume you're the alpha male in the tribe, the one responsible for "bringing home the bacon," if you will. This is quite a task. After all, it's not like you have the ability to pop open the fridge whenever you're hungry. Nothing less than the survival of your entire tribe is dependent upon you (and perhaps your hunting buddies) successfully hunting an animal for dinner. While you're doing this, the women and children are out gathering fruits and vegetables; firewood; and materials for your tools, clothes, and shelter, assuming vegetation is available.

As you scan the plains for a suitable kill, you notice what looks like a bison in the distance. Immediately, you crouch low into the nearby shrubs and ready your spear. Like a lion tracking its prey, you wait patiently in silence to avoid scaring off the animal. This is especially important, as animals have the same fight-or-flight response humans do.

The anticipation of what's about to happen automatically primes your heart to start beating faster. Blood is diverted away from your digestive organs and sent to your muscles. Your body is wonderfully efficient; you're on the hunt, so you don't need to worry about digesting food, right? Your pupils dilate and your vision becomes tunneled, similar to a fighter jet zeroing in on its target. You may even begin to tremble slightly and feel the need to urinate. These are classic signs that your body has entered into the fight-or-flight stress response.

As the bison approaches, it smells your presence. You try to pounce, but it quickly skitters away. Both you and the animal are in full stress mode now. Your sympathetic nervous systems and cortisol and epinephrine levels are sky-high. Both of you have been primed for action. Your body has broken down stored energy (carbohydrates and fat) into fast-acting blood sugar that can now supply the muscles with the necessary fuel to hunt or run. Sadly for you and the tribe, this bison has escaped. What's worse, there are no other potential meals in sight. Hopefully the rest of the tribe was able to gather some plant foods to keep you going in the meantime.

As you can tell, this existence was not easy or luxurious. In those days, this stress response would mean life or death, literally. At this point, however, you might be asking yourself: *How is this relevant to me?*

As we've seen, when faced with stress, your body either readies itself for battle or turns on the afterburners to run as fast as possible in the opposite direction. And that's important in the very short term—it could mean life or death.

The danger, though, is that when your body is continually subjected to stress, its ability to deal with it wears down. The more vigorously and frequently stress calls your adrenals into action, the quicker they will burn out. The more inflammation there is inside your body (from a poor diet, for instance), the more troubled the communication between your brain and adrenals becomes, and the latter may start to malfunction. And remember, since nothing happens independently of anything else inside the body, those worn-down adrenals can then affect other key glands and organs, like your thyroid.

What's the problem with all of this, you might ask?

Well, your adrenals are responsible for producing and secreting some pretty important hormones that are essential to keep your body functioning. Here are just a few:

- **Cortisol:** In moderate amounts, this hormone helps your cells deal with stress.
- **Epinephrine**: Commonly known as adrenaline, this helps ready your body for "fight or flight."
- **Aldosterone**: This hormone plays an important role in balancing the amounts of water and sodium in your body.

If these hormones are compromised by faulty adrenals, then your body will have much more difficulty coping with stress. Over time, as the stress builds, your adrenals become weaker and weaker, until one day they enter full-blown fatigue. That's when things start to get ugly.

From Adaptation to Exhaustion

These days, it's generally accepted that chronic stress can pave the way for more serious health conditions. It's a gradually destructive process that the famous endocrinologist Hans Selye christened *general adaptation syndrome,* or GAS. In a nutshell, this is how it works:

1. Alarm reaction. In the first stage of GAS, the body releases epinephrine, cortisol, and a variety of immediate nerve impulses to deal with a perceived stressor (fight-or-flight response). Your muscles tense and your heart beats faster. Next, your breathing and perspiration increase as your eyes dilate and blood is diverted from your digestive system to your working muscles. Once the cause of the stress is removed, your body returns to its eased state.

2. Adaptation phase. If the cause for the stress is not removed, GAS goes to its second stage, called *resistance,* or *adaptation.* This is how your body copes with chronic stress. Astonishingly, this phase can last for years or even decades. Here, your adrenal glands secrete more cortisol and epinephrine to increase blood-sugar levels to sustain energy and raise blood pressure. High cortisol levels are a hallmark feature of this phase.

If this adaptation phase continues for a prolonged period of time without periods of relaxation and rest to counterbalance the stress response, you become prone to fatigue, concentration lapses, irritability, and lethargy as the effort to sustain arousal slides into negative stress.

3. Exhaustion. In this final stage, your body has quite simply run out of juice. Mental, physical, and emotional resources suffer heavily. Your body experiences a feeling of exhaustion because its coping mechanisms (that is, adrenal glands) have been exhausted. Unlike the previous stage, you're experiencing low cortisol levels because it has been completely tapped out. Now, fast-acting epinephrine overcompensates, and your blood-sugar levels become difficult to stabilize.

Whether this situation arises from overworked adrenal glands or a combination of other factors that impair communication between your brain and your adrenals, the end result is the same—low cortisol, impaired blood-glucose tolerance, weakened thyroid function, and an overall inability to deal with stress. The result? You feel tired almost all the time.

If you weren't tired all of the time, you probably wouldn't have picked up this book, so I'm going to assume that you're most likely now into the *exhaustion* stage. The only way to know for sure is to get tested. I'll show you how in just a moment.

Acute vs. Chronic

Acute conditions and stress are severe and sudden in onset. This could describe anything from a broken bone to an asthma attack or the reaction to a scary moment in a movie.

A chronic condition, by contrast, is a long-developing syndrome, whereby acute stressors or problems persist for an extended period of time. Chronic conditions are the most problematic and are represented prominently by the top four health conditions that plague our modern world—cardiovascular disease, cancer, diabetes, and obesity.

Exhaustion: The Effects of Low Cortisol

The low cortisol levels you experience during the exhaustion stage don't just leave you perpetually run-down, they also make you a candidate for all kinds of health issues. Plain and simple, low cortisol levels can leave you a mess.

One of the first things that starts slipping during this phase is homeostasis. It's the point of equilibrium within the body where everything is working just as it should. Once this goes, substances called *inflammatory cytokines* start wreaking havoc on your body, making you susceptible to inflammatory diseases, mood disorders, skin disorders, chronic fatigue syndrome, chronic pain syndromes, obesity, glucose dysregulation, and even fibromyalgia.[2, 3]

Furthermore, due to down-regulation of the adaptive (humoral) immune system, those with low cortisol become more vulnerable to assaults by infectious and environmental pathogens such as parasites, allergens, certain bacteria, and toxins.[4]

When it comes down to it, you can further identify the low-cortisol state by the following three overriding symptoms:

- High stress sensitivity
- Chronic fatigue
- Chronic pain

Interestingly, you may have heard from experts in the weight-loss industry that cortisol is the devil because it contributes to increased abdominal fat. Yes, this is true, assuming you're in a chronically elevated-cortisol state (the adaptive phase). However, in my experience, most people older than 30 are well beyond that stage and are suffering from overworked adrenals, low cortisol levels, impaired blood-sugar control, and depressed thyroid function—all hallmarks of the dangerous exhaustion stage. In this case, belly fat is the least of your concerns.

Thyroid—Metabolism Master Gland

It's not unusual to hear people attributing their trouble losing weight to their thyroid. It seems to be blamed more and more these days as the weight-loss industry continues to steamroller through our modern culture and many of us try to self-diagnose the health problems we're facing. Although a compromised thyroid can definitely be a factor in your ability to lose weight, it's often indicative of other problems you're facing. In fact, it might also be why you're experiencing low energy.

The thyroid is located around the Adam's apple in your throat. It influences the growth and rate of function of many systems in your body by producing and secreting the thyroid hormones T_3 (triiodothyronine) and T_4 (thyroxine). This process plays a massive role in protein creation and controls how sensitive your body is to other hormones. It's also how your thyroid regulates the temperature of your body and how it uses energy.

Clearly, your thyroid is of crucial importance to many of your body's functions. Unfortunately, it can be damaged very easily. For example, the thyroid requires iodine for the production of T_3 and T_4, but this mineral happens to be one of the most deficient in our food supply due to the stripped, malnourished soils that much of our modern produce is grown in. This is partly responsible for the increasingly common condition of hypothyroidism (low thyroid function), which can also be caused by everything from mercury toxicity to increasing exposure to free-radical damage and radiation (including x-rays and body scanners). If you ever feel sluggish; have cold hands and feet, brittle nails, or flaky hair and skin; or have difficulty losing weight, then you might be suffering from low thyroid function.

Further complicating things is the fact that your adrenal glands also compete with your thyroid for the amino acid tyrosine. What this means is that if your adrenals are shot, then your thyroid is likely underperforming as well, and vice versa. This tug-of-war for tyrosine can in turn make your energy take a huge nosedive.

As you can see, very little in the body happens independently of anything else, but as I always say, how you heal anything is how you heal everything. As we work on getting your energy levels back to where they should be, we'll be working on your thyroid as well.

Adrenal Depleters

You may still be wondering just what you could have done to bring yourself to this point. Sadly, it's probably nothing more than your everyday life. As I alluded to before, the most notable adrenal and energy depleters are dead foods and poor lifestyle factors. Whether you're worrying about paying your bills, consistently engaging in insanely intense workouts, downing coffee or energy drinks, or throwing back a few sugary treats every day, you're "stressing" your adrenals. Because of that, they respond the same way they did 10,000 years ago when your ancestors relied on hunting for survival.

One of the challenges today is that while the human body evolves linearly, technology increases exponentially (according to Moore's law). Thus, we have the same biology we did hundreds of thousands of years ago, which isn't necessarily best suited for handling the increasingly rapid, nonstop stresses of the modern world.

So, collectively, here's how the aforementioned stressors negatively affect your adrenals (and by default drain your energy over time):

Excessive stimulation and stress cause your adrenals to release adrenaline and cortisol—two *catabolic* hormones, which break down your energy reserves for immediate use. Remember, your body thinks it's in a state where it needs to fight or flee—so it needs that readily available energy. When chronically elevated in your bloodstream, cortisol is a detrimental hormone, leading to quick glycogen (stored carbohydrates) depletion, a rise in blood sugar, and eventually weight gain (especially abdominal weight gain). And even when it's less present (adrenal fatigue), epinephrine picks up the slack, yielding the same problems.

Adrenaline (epinephrine) has similar energy-depleting effects. Initially, however, it gives you that feeling often described as a "natural high" or "caffeine rush," because it temporarily increases your heart rate, breathing rate, blood pressure, and carbohydrate breakdown into blood sugar. Within a few minutes you feel alive and ready to take on the world. But that quick surge in energy is soon followed by a debilitating crash in which your body does its best to recover from the stressful situation, whether it's your morning coffee or a near fender bender on your way to work. Can you remember the last time you felt that crash? It probably wasn't too long ago.

Having these two hormones chronically circulating in your blood is not a desirable situation, but this only occurs up until the threshold where your adrenal glands can no longer keep up. Once your body reaches that point, there is less and less hormonal output.

With respect to hormones, balance is key. We don't want excessively elevated levels (that is, acute stress), nor do we want excessively low levels (chronic stress, leading to adrenal fatigue).

If you ever feel exhausted after an argument, a bout of physical exercise, a stressful day at work, or a few hours after your morning coffee, then you can be sure that your adrenals are crying out for help. Everything you feel is a message from your body. In this case, your adrenals are asking for some sorely needed rest and relaxation. They don't want more stimulation, and they certainly don't want to keep up with your intense workout regimen. They simply can't.

As we go through this book together, you'll discover some terrific strategies for reviving and recharging your adrenals and other systems so that you can dump these irritating symptoms once and for all and get on with your life. Right now, let's figure out what condition your adrenals are in.

How to Test Your Adrenals

The gold standard for testing the health of your adrenals is the adrenal salivary hormone test. This test requires four saliva samples to be taken at specific times throughout the day (upon waking, lunch, midafternoon, and before bed) to evaluate your total cortisol level, cortisol (or circadian) rhythm, and DHEA (a precursor to most sex hormones). This is because normal cortisol rhythm during the day features high cortisol upon waking and lowering levels through the rest of the day until bed. Basically, cortisol rises and falls with the sun and is inversely related to the sleep-inducing neurotransmitter melatonin (which does the opposite).

You can get this test done for about $200 at a naturopathic clinic near you or even online through Metametrix Clinical Laboratory (metametrix.com) and Diagnos-Techs (diagnostechs.com).

Aside from lab testing, there are two very simple adrenal tests you can do at no cost right in the comfort of your own home. These will give you a good indication of whether or not your adrenals are working properly.

The first test is called the pupillary light reflex test. Here's how to do it:

1. Stand in front of a mirror in a dark room.
2. Take a flashlight and shine the light into one eye at a 45-degree angle from the side.
3. Watch your pupil for 30 seconds.

When in the dark, your pupil should naturally dilate to allow more light to enter. When you shine the light, it should instinctively contract. The duration of the contraction can give you an indication of adrenal function.

Here's what your findings could represent:

Pupil Constriction	Possible Meaning
Stays constricted for more than 20 seconds	Healthy adrenal function
Pulses after ten seconds	Decent adrenal function
Pulses after five seconds	Poor adrenal function
Immediate pulsation and dilation	Adrenal exhaustion

If negative responses are observed, additional testing is warranted, and I recommend you perform the orthostatic hypotension test. This test does require a blood-pressure cuff. If one isn't accessible to you, I'll share a shortcut with you at the end. Nonetheless, here's how to do it:

1. Lie down and relax for five minutes, then record your blood pressure. Make note of the systolic pressure (the top number).
2. Stand up and take your blood pressure again.

If your systolic pressure remains the same or if it decreases, that usually implies that your adrenals aren't functioning properly. When you stand, epinephrine should be secreted to increase your blood pressure. This helps prevent gravity from pulling blood away from your brain. If your blood pressure drops upon standing, this indicates that epinephrine isn't present to do its job, and that's because your adrenal glands are too tired to work optimally. Make sense?

Now for that shortcut I promised you, as you probably don't have a blood-pressure cuff at home. (Don't worry; I don't either.) The easiest way to perform this test is to simply go from lying down to standing and notice how you feel. Do you feel lightheaded or dizzy? Maybe even like blacking out? If so, these are indicators that your blood pressure has dropped.

There you go—two simple tests that can immediately give you some valuable feedback. If you're ready to take things one step further, then I encourage you to fill out "The All-Day Energy Diet Exhaustion Assessment" that follows.

The All-Day Energy Diet Exhaustion Assessment

The following assessment features 20 questions that relate to symptoms, food patterns, and life events that have affected how you have felt up until today. Answer each as honestly as possible. For each statement, respond with one of the following numbers: 0 = never or rarely, 1 = occasionally or slightly, 2 = moderately intense or frequently, 3 = intense/severe or very frequently. At the end, I'll show you what your score represents.

Before answering these statements, I'd like you to first state when you first started not feeling well and any event associated with that time.

I have been feeling tired and lethargic since (date) ________________ when (event, if any) ________________.

Score	Events, Lifestyle Factors, Food Patterns, etc.
	I have experienced long periods of stress.
	I overwork myself, with little play or relaxation.
	I have used extensive amounts of corticosteroids (creams, inhalers, etc.).
	Lately I seem to get sick more than usual.
	My ability to handle stress and pressure has decreased.
	I am less focused and have trouble concentrating.
	I have trouble remembering recent events.
	My sex drive is noticeably lower.
	I am chronically tired, and it is not relieved by sleep.
	My muscles feel weaker than they should and are quick to tire during exercise.
	I have difficulty getting up in the morning.
	I often have low energy in the afternoon.
	I rely on coffee or stimulants to get me going in the morning and keep me going throughout the day.
	I often crave salt.
	I seem to urinate far too often during the day, especially after drinking water.
	I feel worse if I skip a meal.
	I restrict my salt intake.
	I lose my temper easily.
	I feel exhausted after emotional upset.
	I feel exhausted after an intense workout.
/60	TOTAL

Interpreting your score:

- If you scored between 0 and 20, you don't have much to worry about.
- If you scored between 21 and 40, your adrenals are hurting and need some attention.
- If you scored between 41 and 60, you've got full-blown adrenal fatigue and need to take immediate action.

How did you fare? No matter how you scored, the strategies you'll pick up throughout this book will reduce your scores over time and dramatically improve your energy levels. And don't worry if it seems like a challenge. You're more than capable of pulling this off, and the results can be noticed in just a few days' time. Life only gets more fantastic from this point forward.

■ ■ ■ ■

Chapter Two

3 "SNEAKY" FOODS THAT ROB YOUR ENERGY

Now that you have an idea why your natural energy has been so thoroughly extinguished, it's time to get to work on bringing you back to life.

In the second section of the book, we're going to dive into how to get your body firing on all cylinders again so you can experience the energy you've been so sorely missing. Before you can tackle that, however, you need to start deconstructing and eliminating everything you do that has brought you to this point. That process begins with a closer look at three common foods that rob you of your energy. You might be surprised that they can have such a drastic impact on how you feel every day, but once you start reducing how much of them you eat, you won't be able to argue with how great you begin to feel.

Wheat (Gluten)

There's a good chance you've heard the term *gluten-free* recently. It's echoing everywhere these days. Maybe you have a friend who has sworn to steer clear of the stuff, or you might have seen an article about it in a magazine. Perhaps you've simply noticed more and more "gluten-free" products popping up on the shelves of your local supermarket with each passing month. But what *is* gluten, anyway? Is it really that bad?

Gluten is a protein found in many grains—namely, wheat, barley, and rye. Doesn't sound too bad, does it? Unfortunately, it has now been linked to a number of autoimmune disorders, including celiac disease, a condition in which the absorptive surface of the small intestine is damaged, resulting in an inability of the body to absorb the vital nutrients from the food it digests.

Although only about one percent of the population has actually been diagnosed with celiac disease, the reality is that the human body simply has not evolved to digest wheat/gluten properly. That's right—that buttered toast you ate this morning isn't playing nice with your intestines right now, even if you don't realize it. Chances are, the energy drain you've been experiencing is in some way connected to the gluten you've been eating, as any food that compromises your digestive and intestinal health (most of your nutrients are absorbed in your intestines) is bound to suck your energy.

Continued irritation of the small intestines (by eating grains, for example) can lead to "leaky gut"—a condition in which large, undigested food particles are able to pass into the bloodstream via ever-widening pores in the intestinal wall. When this happens repeatedly, your body identifies the undigested proteins in these food particles as a threat and can mount an immune response to neutralize them. This is the genesis of most food sensitivities and allergies. It's also how your immune system can become overloaded and eventually go haywire. This is actually what happened to me when I developed alopecia.

Why are glutenous grains so disruptive to our health and energy levels? How could bread and pasta be so problematic? The leading theory, which makes a heck of a lot of sense, relates to grains' relatively recent introduction into our food supply.

You see, grains first became a part of the human diet during the agricultural revolution about 10,000 years ago. That's only yesterday when you consider that our genus had been on the planet for roughly 2.5 million years before then. As *Homo sapiens,* we have such little history with eating grain. In fact, grain consumption only represents 0.004 percent of our evolutionary timeline!

This tiny fragment hasn't given the human body enough time to evolve to digest grains efficiently. That's just one reason why so many of us have issues digesting wheat and other gluten-containing grains.

The second reason is that—by definition—grain is a seed. The biological role of a seed is nothing more or less than propagating its species. That's how plants continue to flourish. They spread their seeds, which eventually end up in the soil, where they germinate into full-grown plants or trees, which in turn bear more fruits and seeds . . . and the cycle continues.

However, in order to survive in harsh environments, seeds contain built-in protective mechanisms—anti-nutrients, if you will. Wheat contains gluten, which attacks our intestinal tract, making us unable to digest it properly. Theoretically, gluten's ability to defend itself (and the seed) should allow the seed to pass through the body, into the stool, and back into the earth, allowing the seed's species to carry on. Our delicate gut lining is damaged in the process.

Of course, it's hard for most people to "digest" this information. After all, wheat is such a regular part of the modern diet. This isn't helped by wheat-pushing lobbyists who promote the popular misconception that we need grains for their fiber content, as it improves cardiovascular and intestinal health. The truth is, you can find much better sources of fiber in other foods.

If you want more fiber (and you should—to the tune of about 35 grams per day), there's plenty to be had by eating more vegetables, fruit, and nuts. In fact, Paleolithic humans never ate grains, and their reported fiber intake was between 100 and 150 grams per day from consumption of those aforementioned foods.[1] Our current average intake of 15 grams of fiber per day pales in comparison—even with all those grains in our food supply.[2] The whole-wheat and seven-grain bread you're eating simply isn't all it's cracked up to be.

Sadly, there really are no redeeming qualities, other than taste and convenience, when it comes to eating wheat and gluten. The problem is that the more wheat you eat, the more addicted you become since it contains opioid-like properties. It may seem drastic to cut any and all wheat out of your diet, or even reduce it, but your body will thank you for it. In my work, I've never met a single person whose energy and health has not dramatically improved upon going wheat/gluten-free. You'll feel 1,000 times better.

On top of all this, grains, especially refined ones, wreak havoc on your blood-sugar levels. That's a surefire way to drain your energy. We'll discuss this in more detail as we take a look at how troublesome sugar can be.

Sugar

Everybody knows that too much sugar isn't good for you, but do you know *why?* Sure, it's common knowledge that excess sugar can cause cavities or even hyperactivity, but most of us lack a basic understanding of how sugar works in the human body and just how harmful it can be. It's time for that to change, as sugar is one of the most sinister villains in our battle to regain our lost energy.

You probably know that you already have sugar in your body. It's called *blood glucose,* and it's the first and quickest source of fuel for your brain, vital organs, and muscles. When your blood sugar is too low, you feel "spaced-out," light-headed, and tired, and in some cases you might turn into a sugar monster seeking a quick fix to come back to life. On the flip side, excessively high blood sugar—which you might experience after a triple-scoop sundae smothered with hot fudge sauce, brownie chunks, and candy sprinkles—can leave you feeling "hung-over" and lethargic. Interestingly, the symptoms of low blood sugar and high blood sugar may often feel similar.

For just this reason alone, it's important to moderate how much sugar you consume. You don't want to be stuck on that roller coaster, but it can be difficult, as your body processes sugar from foods not even commonly thought of as sugary. What you might not realize is that when you eat something that is predominantly carbohydrate based (for example, bread, cereal, or pasta), that food is digested and broken down into individual glucose molecules, which then enter your bloodstream. The more refined the carbohydrate (white bread, rather than whole-wheat bread), the quicker the surge in blood sugar.

Now, your body has a natural mechanism that removes excess sugar from your blood. That would be the hormone insulin, which is produced and secreted by your pancreas. It rises and falls along with your blood-sugar levels. For example, if there's way too much glucose in the blood at once, a huge surge of insulin will be secreted to meet the demand. If that's the case, why should you even worry about how much sugar you're consuming?

If insulin has to remove an excessive amount of glucose out of your blood, you experience the sugar crash I mentioned earlier—those awful feelings of being dazed, acting cranky, and hankering for more sugar. The more often this happens, the worse the problem becomes. Eventually, this blood-sugar imbalance turns into hypoglycemia, which is characterized by dreadful blood-sugar crashes within a short time after carbohydrate consumption.

But wait—it gets worse.

If these blood-sugar ups and downs continue over time, your body eventually rebels and says "enough is enough." It soon becomes insulin resistant, and that, my friends, is called *type 2 diabetes.*

Hopefully, you're nowhere near this stage, but you may be caught in the Ping-Pong pattern of sugar highs and lows. Consider the following scenario:

It's 7 A.M. and your alarm clock jolts you out of bed. You feel sluggish and wonder how "sleepy time" is already over. You feel unrested even though you've slept the recommended eight hours. You do your morning business, brush your teeth, and make your way to the kitchen on autopilot. You open your pantry to grab your favorite cereal and pour yourself a glass of orange juice. Things are already off to a bad start.

Why? Well, in a small eight-ounce (240 mL) glass of OJ, you're ingesting about 24 grams of sugar. Furthermore, unless your juice was made from fresh-squeezed oranges just moments prior, you can bet it has been pasteurized or made from concentrate, and is thus devoid of its inherent nutrients. What's left is a tasty glass of orange sugar water.

Hopefully your cereal is a little healthier. Although some cereals on the market are better than others, many of them (especially the kids' ones) are loaded with buckets of sugar. I wonder why so many kids have attention deficit disorder nowadays. (*Hint:* my rule of thumb is to avoid any packaged cereal, or any food product, for that matter, that is advertised on television.)

The average American takes in 22 teaspoons of added sugar a day, which amounts to an extra 350 calories.[3] Some of this we add ourselves. Most, though, comes from processed and prepared foods, including supposedly healthy cereals.

For example, two servings (1.5 cups) of Fruity Pebbles cereal (the amount to fill a small bowl) contain 18 grams of sugar. I'm hoping you and your kids aren't eating this notoriously unhealthy cereal. But what about some of the so-called healthier cereals, like Honey Nut Cheerios or Raisin Bran?

The reality is that these ones are far worse! While Honey Nut Cheerios has an equivalent amount of sugar to the Fruity Pebbles, Raisin Bran has even more, at about 27 grams for 1.5 cups' worth of cereal. Do you see where I'm going with this? This is the typical North American breakfast that we've been indoctrinated to adopt for more than 50 years. Thank you, cereal companies!

It's not even 8 A.M. and between your morning OJ and your "healthy" cereal, you've already ingested about 55 grams of sugar (220 calories), with very little protein, fiber, or healthy fats to mitigate the resulting effects on your now-sky-high blood-sugar levels.

As you get ready for work, insulin starts to be released from the pancreas to usher some of that sugar (glucose) out of the blood and into the cells of your muscles, liver, and fat tissue. And it does a great job, but within a few minutes you're faced with a whole new dilemma—your blood-sugar levels have now fallen through the floor thanks to insulin's work.

Remember, the greater the rise in blood sugar, the greater the release of insulin. The more insulin that is poured into the blood, the more sugar it will remove. The result—far less sugar in your blood than what is ideal. You've now slipped into the realm of hypoglycemia (or low blood sugar). Now you're downright miserable and craving even more sugar and carbs to bring your blood sugar back to normal.

You're pressed for time, so you certainly don't have a few minutes to spare to make anything else at home. As such, you jump in your car, speed off to work, and make a pit stop at your local Starbucks or neighborhood coffee shop to grab a muffin and coffee to really get you going. It does the trick, but in about one hour, you're going to feel like crap again. Worst of all, you still have a whole day of work ahead of you.

Can you see how this vicious cycle can easily get out of control? What ends up happening to many people is that they simply bounce from one sugar or caffeine high to the next: morning cereal to muffin to bagel to sandwich to midafternoon chocolate and on and on and on.

The solution is to become mindful of the amount of sugar in everything you eat. That also goes for cereal, muffins, and bagels, which you've probably been led to believe are good for your health. Not only are they filled with gluten, they're also high-glycemic-index foods, meaning they're quickly digested and cause sharp spikes in your blood sugar. Whichever form you eat or drink it in, sugar is a stimulant, which inevitably forces your adrenal glands and pancreas to pump out stress hormones and insulin, respectively. As sweet as it may be, there's a good chance sugar is sucking the energy out of your body, accelerating your aging process, and slowly but surely robbing you of your health.

Caffeine

So you don't feel like jumping out of bed in the morning. And you find it tough to focus after lunch as you enter that afternoon lull. Don't worry—you're not alone.

Every single morning, hundreds of millions of people crawl out of bed and turn on their coffeemakers immediately after they turn off their alarm clocks. Others prefer to grab their cup of joe on the way to work. If you're a citizen of Caffeination Nation, then I invite you to seek out citizenship elsewhere; otherwise, by relying on caffeine, you are essentially borrowing energy from tomorrow today. Look at what that type of behavior has done to consumer debt in many nations and to the economy in general. That's what you're doing to your body with caffeine.

The first and most important thing to remember about caffeine is that it's a stimulant, and what do stimulants do? They *stimulate* your adrenal glands into that fight-or-flight response we've already discussed, prompting a surge of cortisol—if you even have any left—and epinephrine into your blood. Since these stress hormones break down energy reserves, they inevitably cause a rise in blood sugar. We've just seen how damaging that can be.

The more caffeine you consume, the more this scenario plays out. Add sugar to the mix and you've just created a crash cocktail. The problem is that most people don't consider the downside of these stimulants, nor do they reflect on the fact that the body metabolizes them the same as any food. They only attribute the immediate high they feel to the coffee or energy drink they just ingested. But don't forget—what goes up must come down, especially when stimulants are involved.

The initial feeling of mental alertness and energy that a hot cup of coffee brings on is almost always followed by a dreadful crash as your blood sugar plummets and your adrenals do their best to recover from another bout of acute stress. For most of us, the "solution" is another cup of coffee, or maybe a caffeinated soda or energy drink. It too is followed by another crash. By the end of the day, you've been up and down on this seesaw quite a few times. If this process repeats itself day in and day out, it's only a matter of time before your body breaks down. There's no amount of coffee that can perk you up from that.

Caffeine simply masks what's really going on inside our bodies and temporarily gives us a false sense of feeling alert and productive. It's a Band-Aid solution that will never give us the lasting energy we crave. In fact, over time it just makes things worse.

Our addiction to caffeine points to a sad but unavoidable truth: it's nothing more than a drug. In fact, it might be the most commonly accepted drug in the world. As with any drug, habituation and dependency become the norm rapidly.

Habituation means that your body quickly adapts to the dose of the drug, which means that you need more of the drug to experience its desired effects. Dependency refers to the fact that you depend on the drug to function normally. If you currently consume caffeine, in any form, try going caffeine-free for a day or two. You'll quickly feel like crawling into a hole. You'll be in the throes of a full-on caffeine withdrawal, as the caffeine receptors throughout your body are no longer being satiated, so you feel terrible. It's the same type of withdrawal experienced by those trying to give up nicotine, alcohol, heroin, or any other drug. Physiology doesn't judge—a drug is a drug is a drug.

Let me be perfectly honest with you—nowadays I might have two to three decaf lattes a week at most. I don't drink caffeinated beverages because I know how caffeine negatively impacts my body, but I wasn't always this way. I actually started drinking coffee when I was 23 and living in France. Being able to walk to the local café and grab an espresso or *café crème* was something I somehow fell into, probably due to the influence of some of my soccer teammates. Nonetheless, I really enjoyed those moments and remember obsessing about my next espresso. At the time, I hadn't switched to decaf, so I was experiencing caffeine's full effects—at least initially.

It wasn't long before my daily espresso wasn't enough. I was no longer experiencing the "high" I had previously, so I upgraded to a double espresso and started making more trips to my local café throughout the day. After moving back to Canada and switching over to lattes and cappuccinos, I started to notice that I felt more tired *after* my caffeine fix than before it. I didn't know why. Thankfully, I do now.

In fact, not too long ago, years after I finally gave up caffeine for good, I remember going to Starbucks to do some work before an evening soccer game. I ordered a decaf latte and spent an hour working before heading to my game. As I got on the field and started my warm-up, I was shocked by how "out of it" I felt. I was jittery, sweaty, and really anxious. At that point, I realized that the barista must not have heard that I had asked for a "decaf." I felt absolutely terrible. Not surprisingly, I didn't have a very good game and I crashed hard when it was done.

I have to admit, I really do enjoy the taste of a good coffee. Sometimes I'll even go out of my way to get one. The thing is, I'm not after the buzz but rather the flavor and the experience. Emulating the "world's most interesting man" commercials, let me finish this story by saying this: "I don't always drink coffee, but when I do, it's decaf."

Now, I've had numerous discussions with friends who consider themselves true coffee aficionados. They've studied coffee beans from around the world and are quite familiar with some of the finest blends you can find. Coffee, in its caffeinated form, fuels their performance and productivity, and they can't imagine their world without it. They've tried to convince me that certain beans contain different types of caffeine that are gentler and longer lasting in the body. As a result, they don't produce crashes and negative consequences.

As I see it, that's like saying that one particular brand of scotch contains only 15 percent alcohol while another contains 20 percent. Alcohol is alcohol, and caffeine is caffeine. If you have caffeine coursing through your veins, its effects will be fairly consistent regardless of the source. That's not something I want, and if you're struggling to keep your energy levels up, it's not something you should want either.

Hopefully, you're getting that seriously reducing your intake of stimulants like caffeine is imperative to good health and lasting energy. To begin your rehabilitation, make the switch to decaf coffee. As we continue through this journey together, I'll show you alternatives to your morning coffee that will skyrocket your energy—without the crashes—and have you questioning why you ever relied on caffeine in the first place.

■ ■

Even if you aren't a coffee drinker, I want to bring to your attention how much caffeine and sugar is actually present in many of the beverages you might be drinking right now. First, it's helpful to realize how much caffeine is considered acceptable or safe. Shockingly, Health Canada recommends not exceeding 400 mg (0.014 oz) of caffeine per day. As you'll see below, this amount equals about three eight-ounce (237 mL) cups of brewed coffee per day! According to this logic, it's perfectly fine for teenagers to continue guzzling two big cans of Rockstar (about 360 mg caffeine) each day. As a holistic nutritionist, father, and logical person, I find this recommendation absolutely absurd.

In spite of Health Canada's guidelines, most medical boards advise an upper limit of 300 mg of caffeine per day.[4] These levels will still produce deleterious effects over time.

Doing my own research (that is, looking at beverage nutrition labels), I've compiled the following table to show you how much caffeine and sugar is present in many popular beverages. I hope this is enlightening.

Beverage	Size	Caffeine	Sugar
Coca-Cola	12 oz/355 mL can	35 mg	46 g
Snapple Lemon Iced Tea	16 oz/473 mL bottle	43 mg	23 g
Starbucks Latte	16 oz/473 mL (grande)	150 mg	N/A

Starbucks Drip Coffee	16 oz/473 mL (grande)	330 mg	N/A
Tazo Chai Tea	8 oz/237 mL	47 mg	N/A
Rockstar Energy Drink	16 oz/240 mL can	160 mg	31 g
Red Bull	8.4 oz/250 mL can	80 mg	27 g
5 Hour Energy	2 oz shot	138 mg	N/A

As you can see, it doesn't require much to reach the upper limit of 300 mg of caffeine per day. Unfortunately, most people who regularly down energy drinks are already there and require even more caffeine to start feeling its effects. It's downright scary.

Along with gluten and sugar, caffeine forms an unholy trinity of substances that are ravaging us all. In fact, they're some of the primary culprits that are responsible for the personal energy crisis so many of us are facing.

By eliminating them from your diet, or even cutting back a little, you'll be taking a tremendous step toward your triumphant return. Give it a try; by taking this step, you're paving the way for the work to come.

■ ■ ■ ■

Chapter Three

THE UNTOLD SECRET INNER WORKINGS OF YOUR BODY

Without electricity, you can't turn on the lights in your home. Similarly, without gasoline, your car won't start or go anywhere. So what is it that allows your body to perform basic functions such as lifting this book (or your e-reader) and reading this sentence?

It's a pity that we don't gain a better understanding of how our bodies work when we're growing up. DVD players and washing machines come with instruction manuals, so why don't our bodies? I mean, we are taught a few things as teenagers, but how much of it do you actually remember? Can you recall anything about the Krebs cycle? Does the term *glycolysis* ring a bell?

Probably not, but that's okay.

I'm not going to berate you for failing to remember all the details of biology class, but I do think it's important for you to understand your most basic and fundamental inner workings. You're on a quest to become the fully energized being you were born to be, and part of that mission involves knowing what powers your ability to speak, walk, and even think. Simply put, you need to know where your energy comes from.

Don't worry—I'm not dragging you back into the high school science class you should have taken, or even the one you took and simply didn't pay attention to. In fact, we're going to skip the boring stuff and get right to what you need to know, the information that will make a lightbulb pop up over your head.

I firmly believe that understanding where your energy comes from will make it easier for you to overcome the sloth-like days you've been facing. For so long, you've been juiced up on stress and caffeine, thinking that sheer determination will help you through your day, but it's clearly not enough. You wouldn't be reading this book if it was. Once you know what generates energy in your body, it will be that much easier to kick the bad habits that get in the way of this process. Better yet, you'll be able to stoke your internal fire even more and really get your engine going.

Actually, you might want to consider *this* the instruction manual to your body that you never received.

What Really Gets You Going

The first thing you need to understand is that you're nothing more than a bunch of cells.

I hope that doesn't sound like an insult, but it's the absolute truth. You are a you-shaped collection of cells—trillions of them, in fact. Everything inside and outside of you is made up of these tiny, microscopic things, from each millimeter of your toenails on up to that big, marvelous heart of yours.

Now you've probably heard the term *cell* used in conjunction with something else before: batteries. That's because the cell of a battery is essentially its heart, the very source of the energy it needs to operate and power whatever you put it into.

Your cells serve the same function inside of you. They power you up.

Now we could get into all the intricacies of how your cells work, but I promised not to bore you, and I intend to keep that promise. What you need to know is this: your cells create energy using something called *adenosine triphosphate,* or ATP for short.

As complex as that sounds, your cells only need two things to create it: glucose from the food you eat and oxygen from the air you breathe.

That's right: all of your energy comes from eating and breathing.

Now that sounds very simple, and it really should be, but the way we live these days complicates matters. As we've discussed, most of us eat far more than we need to, which means our cells are getting more than enough glucose to create the ATP they need. For those of us struggling to get by with low energy, what our trillions of little cells are so sorely in need of is more oxygen.

That might sound a bit odd given the fact that you're breathing right now. If you weren't, you wouldn't be alive. Therefore, there's a simple question that needs to be answered: if you're breathing about 12 to 18 times per minute, as the average adult should, how come your cells aren't getting enough oxygen to produce ATP?

The answer lies in your blood.

Why Your Blood Is So Important

It's quite all right to admit that you don't know what specific purpose your blood serves. The mere thought of it might even scare you. After all, the only time you see it is when you're hurt in some way.

That said, I'm quite certain you know you can't live without it, even if you don't know exactly why. Much of it has to do with the oxygen your cells need.

Your bloodstream is a river of life, transporting oxygen to the cells that need it via your red blood cells. The whole process begins the minute you breathe. With each inhalation, air flows down to little air sacs at the bottom of your lungs called the *alveoli*. Here, the oxygen you just inhaled hitches a ride on your red blood cells almost as if it was taking a cab. These red blood cells then take it wherever it's needed, be it the cells in your heart, your brain, or any other organ.

Pretty straightforward, right? You breathe, and the oxygen is quickly transported via your blood all over your body. That's a simple enough process.

However, like any transportation route, there can be delays and slowdowns. If you've ever taken a taxi in New York City during rush hour—or even if you've only seen it on TV—then you know what it's like to be stuck in gridlock, bumper-to-bumper traffic: nothing moves. The same thing can happen in your bloodstream. That's what's happening to so many of us, and it's exactly what was happening to me for such a long time. When your red blood cells aren't flowing as quickly and easily as they should, the cells throughout your body don't receive the oxygen they need when they need it, and are unable to produce energy. As a result, you feel horrible and completely deprived of energy.

This dreaded traffic jam in your bloodstream occurs when your red blood cells are clumped together. How do they get that way? It all has to do with the acidity of your blood, and that's caused by the food you eat.

Do you see how everything in your body is interconnected?

Acid vs. Alkaline

The pH scale is probably the last thing that comes to mind when you think about your blood. Why does it matter if your blood is acidic or alkaline?

Ideally, you want the pH of your blood to be slightly alkaline, just over the middle of the pH scale, with a measurement of 7.35 to 7.45. Here, your blood is considered to be in homeostasis, which is the ideal point of balance that allows everything to function correctly. In homeostasis, your red blood cells are fully formed and circulating freely all over your body. There are no traffic jams of any sort.

Unfortunately, much of our modern diet causes our blood to tip toward the acidic end of the pH scale, and this causes trouble for those regularly scheduled oxygen deliveries our cells rely on to produce energy. Meats, dairy, processed grains, and sugars are all acidic foods, and eating too many of them turns our blood acidic. Sadly, those foods make up much of what we tend to eat. Junk foods like cheeseburgers and ice cream are entirely made up of these acidic food types, but so are seemingly less troublesome foods like chicken alfredo and ham-and-cheese sandwiches.

Come to think of it, there's a good chance that most of what you ate today was composed of acidic foods.

Again, your cells are like little batteries; more specifically, your red blood cells have a positive and a negative charge. In an acidic bloodstream, the outer negative charge is stripped from your red blood cells and their inner positive charge becomes exposed. That's a problem, because the basic principles of magnetism apply to red blood cells as well: negative repels negative, and positive repels positive. Furthermore, positive attracts negative.

If you have a bunch of red blood cells floating around only emitting a positive charge, they end up "sticking" to other red blood cells that still have their negatively charged outer shield intact. This is what creates the "traffic jam"—clumps of red blood cells floating around together when they should be zooming all over the place. If your blood is at the proper pH, all of your red blood cells keep this negative charge going, and thus repel each other, allowing your bloodstream to flow smoothly. This, in turn, allows the oxygen you breathe to be delivered efficiently to the organs that need it.

If you want alkaline foods, you have to turn to vegetables and fruits, especially the green ones. We'll get to why the color green is so important later in the book, but for now, just know that vegetables and fruits are your friends. Sadly, most people choose to ignore this sound advice. They eat their veggies out of bare necessity when they should be a major part of anyone's diet. An occasional side salad isn't enough, and the blueberries in the muffin you had this morning don't count, either.

It's often said that most people's diets are about 80 percent acidic and 20 percent alkaline, when the reverse should be what we strive for: 80 percent alkaline foods and 20 percent acidic foods. It's an easy choice, but an uncommon one.

I'm not telling you that you have to run out and become vegetarian, but becoming more mindful of what you eat is essential to getting your energy back on track. It's not something I expect you to understand how to do intuitively, especially when you consider the bad conditioning most of us receive around eating. That's why I've dedicated the All-Day Energy Diet plan in Chapter 6 to laying down a specific diet and schedule to help get you back in balance.

Why Your Doctor May Scoff at This Idea

There's an uncomfortable truth around blood alkalinity, and it's this: for the most part, the modern medical establishment doesn't regard the pH of your blood as an issue worth considering. Talk to your doctor about it, and he or she may scoff at your interest in this "invalid" concept.

Don't let this discourage you. Actually, let's take a look at why your doctor's hubris is worth questioning, and even countering.

As you've seen from my medical journey, I didn't start receiving the answers that I so desperately needed until I left "traditional" medicine behind and steered toward holistic nutritionists. Despite being a sick mess of a man, I just couldn't be healed by the scores of doctors and specialists I sought out looking for solutions. It wasn't until I started consulting some *truly* traditional principles about healing that I was able to bring my body into balance again.

The sad truth is that your average doctor is not schooled in the field of nutrition. M.D.'s can diagnose diseases of all kinds and give you expensive, annoying pills to get rid of your symptoms. Sometimes, they might even suggest surgery. However, it's increasingly rare to find a doctor who can explain the origin of whatever disease or condition you're struggling with.

So often, simple medical solutions elude us because we've become unnecessarily tied to the thinking that a cure is better than prevention. How about solving our health problems by addressing the root causes?

Numerous scientific studies have shown the benefits of an alkaline approach for an arsenal of health issues such as rheumatoid arthritis, kidney disease, obesity, cardiovascular disease, diabetes, acid reflux, gout, loss of bone density, and more![1, 2, 3, 4, 5, 6]

But for some reason the medical establishment wants to tuck these truths under the table.

I don't want to rant, so let me give you an example that pertains specifically to what we're discussing in this chapter: the alkalinity of your blood, and how important that is to your energy levels.

Again, many doctors will scoff at the idea that your blood pH has any impact on your health. Simply ludicrous, they'll say. However, they very likely will say that you need to incorporate more fruits and vegetables into your diet. Why exactly do they offer this suggestion? They probably will make mention of your need for more fiber and minerals, but they'll ignore the fact—although they probably don't know it to begin with—that boosting your vegetable and fruit intake will have a massively positive impact on your blood alkalinity and thus your cell health. You'll be repairing your body on the cellular level. Without realizing it, they're supporting the very thinking they so readily dismiss!

Do you see why I get so frustrated with modern medicine? Now I'm not suggesting you don't go to the doctor, or ignore anything your general practitioner has to say. That would be absurd and irresponsible of me. What I am saying is that there are aspects of your health that your doctor may not be familiar with, because these insights don't support the financial interests of pharmaceutical companies or the health-care industry as a whole. You owe it to yourself to investigate healing modalities that produce tangible effects and provide practical insights. As with anything else, it's all about balance.

How to See If Your Blood Is Healthy

If you're interested in a more thorough investigation of your blood health, then I highly recommend going to a naturopathic clinic to get a live blood cell microscopy done. Essentially, it's a close analysis of your live blood cells.

This is quite different from the blood sample you get done at your general practitioner's office. In that procedure, your blood is taken via syringe, frozen, and sent off to a laboratory. There it's analyzed for its cholesterol and blood-sugar information, which is very crucial, but not what we're aiming to understand here.

In a live blood cell microscopy, a drop of your blood is placed under a very high-powered microscope the minute it is taken from your body. This allows you and your naturopath to look at your blood in real time to see how your red blood cells are behaving. When my naturopath and I took a look at my blood, we saw exactly what so many people would see if they undertook this procedure themselves: a full-on, 5 P.M. traffic jam in my red blood cells. It was just the confirmation I needed that I had to do something about my diet, and it was an important step toward the creation of the very plan that graces the pages of this book.

It's enough to use how you feel as a gauge, and changing your diet for the better is recommended for everyone, even if they aren't experiencing a personal energy crisis. However, if you want to dig deeper and get some specific information on the state of your blood health and how it's affecting your energy, then I definitely recommend looking up some naturopaths near you.

How You Make Matters Worse

I'm sure you've already gathered that I have no love for caffeinated coffee or energy drinks. Here's yet another reason why I'm so vehemently against them: they're acidic.

We've already seen that these two culprits interfere with your cortisol levels and, after they've given you a temporary surge of energy, ultimately end up making you even more tired than you were when you consumed them. What makes them worse still is that they contribute to your blood acidity, wearing you down even more in the long run.

The further we'll get into the book, the more you'll see that so much of what you ingest is bad for you in a number of ways.

Caffeine and high sugar content attack you in many places at the same time, so you're utterly defenseless. That's downright scary!

I have a little more information to share with you before we start improving your diet, but I hope that by the time you get to that chapter, you've already started to wean yourself off of these terrible quick fixes. If you haven't already, I hope this additional information drives home this point: they're just no good for you. It might be difficult at first to give up these energy "crutches" you've been relying on, but give it a few days and you'll naturally start to reset yourself. Then you'll truly be ready for the incredible diet info yet to come.

■ ■ ■ ■

Chapter Four

THE DIRTY DAILY HABIT THAT MAKES YOU SLUGGISH . . . AND HOW TO FIX IT

Here's a little secret: the biggest energy drain you face is tied to something you do every single day. In fact, it's so obvious that you might be overlooking it completely. Have any ideas?

If you're puzzled, then just think back to your last Thanksgiving dinner. Picture the golden brown turkey fresh out of the oven, the fluffy mashed potatoes, and the warm dinner rolls. Is your mouth watering yet? If not, close your eyes and allow yourself to really sink into the memory; time-travel back to that dinner table. Can you taste the succulently sweet cranberry sauce? How about that addictive, savory stuffing you can never get enough of?

Fast-forward to the very last forkful of turkey that goes into your mouth, and I don't mean from your first plate. I mean, you have to have seconds at Thanksgiving, right? How do you feel? I would imagine you're pretty full, but the pumpkin pie is about to go out on the table, and you just have to have a slice even though you're absolutely can't-eat-another-bite stuffed. After all, it's Thanksgiving. Bloat and bellyache be damned!

Now how do you feel? Are you even awake? Chances are, you politely excused yourself from the table, waddled out into the living room, and promptly slipped into a food coma on the couch. You'll probably still be pretty out of it when you wake up, and despite your lethargy, you're going to head into the kitchen to rustle up some leftovers. When you're done, you'll feel even worse.

If only we had a switch in our brains that quickly reminded us of every single time we made this mistake. If only there were some way to remind ourselves that shoveling food down our throats and then somehow finding room for dessert was not a good idea. The thing is, we humans can be a little slow sometimes. We repeat the same behaviors over and over again, even if there's a negative consequence attached to them. The problem in most cases—at least when it comes to food—is that these immediate consequences aren't painful enough to force us to stop.

For example, if you ate turkey and then immediately got a bout of food poisoning, the chances of your wanting turkey again are slim to none. That's a powerful aversion, and your brain will have a tough time forgetting the pain caused by all that vomiting and diarrhea.

But what about feeling tired and sluggish after a meal? Does that create a strong enough aversion to certain foods? The answer is no. That's why we continue to eat way too much of the same foods over and over again in spite of the fact that they may make us pass out as soon as we get up from the table.

I know from experience. There was a time when I was guilty of the most gluttonous kind of Thanksgiving feasting. In fact, many of my memories of the holiday are blurry thanks to the vegetative state I would eat myself into. I'd practically eat whole pumpkin pies by myself!

Over time, however, I came to value how I feel after a meal, and that has meant being more mindful of what I eat and how much of it I'm eating. That's because the act of digesting a meal takes a much bigger toll on our bodies than most of us realize.

Every day may not be Thanksgiving, but ask yourself: how often do you feel an urge to take a nap after you've eaten? It probably happens more often than you'd care to admit.

The sad truth is that most of us are suffering from bad digestion, and it's affecting our health and energy in ways we don't realize.

What I'm about to share with you over the next several pages is very important. In fact, I'm going to mention a few things that will probably challenge all you've ever thought and been told about digestion, food, and energy. All I ask is that you keep an open mind.

How Your Body Digests Food . . . Normally

You're probably not thinking about the intricacies of digestion when you take a bite of pizza. Who can be bothered to consider what happens after you swallow?

I hate to break it to you, but understanding what happens inside of your body when you eat a meal is critical to breaking through the energy slump you're trying to get out of. Once you understand the importance of digestion, you'll be primed to eat in a manner that's consistent with the all-day energy that seems so elusive to you now.

So exactly how important is digestion? Think of it like this: your digestive system is nothing less than your engine. If the engine of your car breaks down, then you aren't going anywhere. In fact, you might have to consider buying another car. Unfortunately, you can't buy another body, so it's important to get your digestion in order. Everything you do depends on it.

Most people don't realize their digestive system is this important, because digesting food is something that we tend to take for granted. We eat and drink, and everything else seems to take care of itself. The little-known truth is that the very foundation of our health begins in our digestive process. It's not common knowledge, but so many of the sicknesses and conditions we face—everything from Alzheimer's to acne—can be linked to improper digestion. In fact, our digestive process is so crucial to the optimal function of our bodies that in medical and scientific circles, it's often referred to as the "second brain."

Despite what you may think, digestion doesn't begin after you've swallowed a bite; the whole process gets under way the minute you put a piece of food into your mouth. Furthermore, chewing doesn't just break down your meal and make it easier to swallow; it also sends a signal down to your stomach that food is on the way. At the same time, the salivary glands in your mouth begin to produce and secrete saliva, and that's when things really get rolling.

Saliva is rich in enzymes that help you break down more basic foods like simple carbohydrates. For example, if you were to leave a piece of white bread on your tongue without chewing it, you'd soon notice the bread dissolving. That's your salivary enzymes getting to work.

Do the same thing with a piece of steak and you'll experience nothing other than a hunk of meat sitting in your mouth. That's because protein requires different enzymes and a more acidic environment in order for it to be digested. That's where your stomach comes into play.

As food leaves your mouth, it travels down your esophagus and into your stomach, a highly acidic environment loaded with hydrochloric acid and enzymes like pepsin, both of which are critical for protein breakdown. Bigger proteins, like meat and eggs, are deconstructed into smaller peptide chains and eventually into individual amino acids as your meal now travels from your stomach into your small intestine.

The small intestine actually isn't that small. In fact, it measures several feet and has the surface area of a tennis court! This is ideal, because the nutrients in the food you eat are absorbed along the walls of this enormous, not-so-small intestine, which provides the ideal environment for most of your nutrient absorption to take place. It's an alkaline environment, and it's here that the digestion of fats, carbohydrates, and proteins is completed thanks to a massive influx of digestive enzymes from your pancreas.

Finally, whatever food has not been absorbed in the small intestine enters into your large intestine (colon), where water and a few leftover nutrients are absorbed. Like the stomach, the colon is more acidic and hosts over 400 different types of microorganisms. This *gut flora* is where your good bacteria (that is, probiotics) keep the bad bacteria, yeast, fungus, and other critters in check. Your gut flora is so important that roughly 80 percent of your entire immune system surrounds the lining of the gut. Known as your *gut-associated lymphoid tissue* (GALT), this is how your body communicates with the outside world (in other words, the gut flora) to keep you safe and healthy. After your meal has been processed by all of these different parts of your digestive system, what remains is a fibrous clump of fecal matter that is to be excreted upon your next bowel movement.

As odd as it may sound, it's important to understand that your digestive tract (from your mouth to your anus) is actually considered *outside* of your body. Only when nutrients are absorbed out of the digestive tract and into your bloodstream have they entered the *inside* of your body. Also, most of your body's primary defense systems are found within this long, winding digestive tube. If something is dangerous, infectious, or too big, it doesn't get in—ideally.

From mouth to anus, food usually takes about 18 to 24 hours to go through this cycle. If it takes longer, then something isn't working properly. This could lead to, or indicate, constipation. If transit time is much shorter, something in your food could be irritating your digestive system, forcing it to literally purge what's inside. This leads to diarrhea or unhealthy bowel movements.

That's a lot of activity, isn't it? It also means there's a lot of room for error, and given the fact that most people in our society are addicted to the SAD (Standard American Diet), many of us don't digest our food anywhere near this efficiently. Sadly, we have "kinks" in our plumbing thanks to the terrible eating and lifestyle habits we overlook. That means our bodies are receiving very little nutrition from the food we eat, and we're also making ourselves vulnerable to disease and sickness.

Before we can start fixing your digestive system, we first have to understand all the problems you might be facing. Let's take a look at where so much of what's inside of us goes wrong.

The Importance of Enzymes

In that little tour I gave you of your digestive process, you may have noticed that enzymes are present both in your mouth and in your stomach and intestinal tract. They can even be found in your food. They're fascinating, versatile little things that are essential to life itself, no matter what form it takes.

Technically, enzymes are proteins that accelerate certain reactions in the body. There are so many different types of them, but they can be separated into three main categories: *digestive enzymes* (ones that work on food digestion), *metabolic enzymes* (involved in every other bodily process), and *food enzymes* (enzymes inherent in living foods). The ones we're most concerned with break down the food we eat. For example, lipase digests fat, while amylase handles carbohydrates. Protease tackles proteins, while delta-6-desaturase works on metabolizing omega-3 and omega-6 fats, and so forth.

As with many things in the body, enzymes can be depleted due to the same factors that tax our adrenals: age, stress, unbalanced diets, and exposure to toxins. It is extremely common for our bodies to no longer produce the necessary amounts of each enzyme to properly digest the food we eat. That's a big problem.

When you don't have enough enzymes to properly digest your food, the end result is impaired health due to digestion-related food intolerances, nutrient deficiencies, gastrointestinal issues and distress, and a whole host of other related problems.

What's really scary is that an enzyme deficiency can still sink your health even if you suddenly start eating healthy foods. Why? Well, you've probably heard the expression "You are what you eat," but the reality is that you actually are what you *absorb:* with an enzyme deficiency, your body is unable to absorb all the nutrients from the food you're eating, even if you're having salads and super-smoothies twice a day. Yup, even by eating the healthiest foods, you might still suffer from food intolerances and nutrient deficiencies that commonly arise from less-than-optimal enzyme production.

Exhausted Enzymes, Food Intolerances, and Weight Gain

If you've lived with a severe food intolerance, you know how much of a pain it can be. Ordering food from a menu is like tiptoeing through a mine field. You may be interested to know that insufficient and exhausted enzyme pathways are two big contributors to how your body processes and responds to the foods you eat. Two of the most common ways your body can develop intolerances, sensitivities, or allergies to foods include:

- Repeated exposure to the same food
- Impaired digestion and intestinal motility, meaning that food sits too long in your digestive tract

Both of these events can deplete your body's enzymes, weaken the important integrity of your gut, and predispose you to allergenic problems—one of which is weight gain.

For example, in one study conducted at Baylor College of Medicine, 98 percent of participants decreased their body fat percentage and/or body weight by simply removing foods from their diet in which they tested positive for intolerances. Meanwhile, the control group who followed a calorie-restricted diet alone (which still included untolerated foods) actually *gained* weight.[1]

More recently, a study presented in the *Middle East Journal of Family Medicine* placed subjects on a diet free from their specific allergenic foods. Over the course of the 12-week study, subjects lost an average of 37 pounds, virtually all of which was body fat.[2]

Thankfully, by ensuring an adequate intake of food-based and digestive enzymes (which we'll look at later) that help your body fully break down and absorb the nutrients contained in any and every food you eat, you can reduce the likelihood of these unwanted food reactions taking place.

The Importance of Gut Flora

As I mentioned before, the very heart of your immune system is in your digestive system, specifically in the gut-associated lymphoid tissue (GALT). It's your first line of defense against enemy invaders—toxins, bad food particles, infections, and pathogens.

This is a deeply misunderstood fact, a misunderstanding that has thwarted our efforts to deal with disease and sickness. We need to understand that we don't just get sick by "catching" an illness; in many instances, our gut flora has become compromised, thus weakening our immune system and making us prime targets for a whole host of health problems.

How exactly does your gut flora become compromised? A bad diet is one of the main culprits, as are antibiotics, which eliminate "bad" gut flora that may be making you susceptible to the illness you're experiencing, but also do away with the "good" gut flora that help you heal and maintain your health.

The Importance of Stomach Acid

Your stomach is hardwired to be very acidic. As we saw earlier, this is extremely important for protein digestion.

Furthermore, the acid in your stomach also serves the purpose of destroying unfriendly bacteria and other pathogens. With less acidity, you become more susceptible to *H. pylori* infections (that is, stomach ulcers), bad breath, and a host of other problems. Stomach acid is critical.

If you're experiencing digestive issues, there's a good chance that what you're battling is low stomach acid. It's a common condition—scientifically named *hypochlorhydria*—and it's probably draining your energy.

When you're suffering with this condition, your low stomach acid struggles to break down proteins like beef, chicken, and eggs so the amino acids and nutrients they contain can be properly absorbed by your body. If this is not able to occur, you'll absorb only a fraction of the protein you're eating, and chunks of undigested food particles can make their way farther down your digestive tract.

Under optimal conditions, all foods are fully broken down into their individual building blocks (that is, glucose, fructose, amino acids, and free fatty acids) and easily absorbed through their specific channels or "pores" in the intestinal tract. However, when larger, these chunks of undigested protein molecules are too big to be absorbed and end up making their way through the intestines. Down there, they can putrefy and give off smelly gases that will "liberate" themselves from your body through awful breath and noxious farts. Worse, they can also bind to your intestinal walls, causing inflammation, irritation, and a subsequent widening of intestinal pores, eventually creating a nasty condition known as *leaky gut syndrome.*

Leaky gut—scientifically named *intestinal permeability*—is characterized by increased penetrability of the intestinal walls, allowing undigested protein molecules into the bloodstream. With these unrecognizable protein molecules traveling through the blood, your body reacts by mounting an immune response to take them out. After all, these food proteins aren't where they're supposed to be, so they're seen as foreign invaders. Your body is simply doing what it knows how to do—fighting off UFOs: *unidentified floating objects.* This entire process is how food sensitivities, allergies, and eventually full-blown autoimmunity develop. There are, of course, other factors, but faulty digestion and intestinal permeability is a big one.

From an energy perspective, if your immune system is constantly working on fighting UFOs and your sluggish stomach is constantly inundated with foods to digest, there's little energy left for you to do anything other than drift through your day like a zombie. Your body is working overtime; of course you feel half-dead half the time!

The Scourge of Acid Reflux

One of the biggest stomach problems so many of us face is acid reflux disease. Sadly, most people don't even realize they have it.

Acid reflux occurs due to the improper functioning of your lower esophageal sphincter, also known as your cardiac sphincter. Its purpose is to keep food and acidic digestive juices in your stomach from washing up into your esophagus. Factors that can increase your risk of acid reflux include smoking; alcohol; bad fats; coffee; obesity; big meals; and drugs, including tricyclic antidepressants, anticholinergics, nitrates, and calcium channel blockers—just to name a few. Many of us suffer with this frustrating condition, but are quick to pass it off as nothing more than heartburn.

Acid reflux can also occur if you have insufficient stomach acid, as the muscles of your stomach are forced to churn even harder. In the process, some of the acid from the stomach may seep back into the esophagus. Additionally, low stomach acid allows unfriendly bacteria to flourish throughout the digestive tract. Normally, the secretion of amylases (the carbohydrate-digesting enzymes) into the small intestine is dependent upon acidic foodstuff leaving the stomach and entering the small intestine. However, with low stomach acid (and thus low presence of acidic food in the small intestine), amylase release is minimal, leaving plenty of carbohydrate for those nasty bacteria to feed on. As a result of their feeding frenzy, fermentation occurs and gas is created. And this gas pushes back into the stomach and can force some of the acid in the stomach back up into the esophagus.

The problem with this troublesome condition is that many doctors prescribe "antacids" to soothe the burning, but as their name implies, antacids can actually make matters worse by turning your stomach more alkaline. If your stomach loses its acidity, it loses its ability to digest food properly. This creates an even bigger problem over time.

Proteins aren't the only things left undigested by low stomach acid. Carbohydrates will also be left to ferment without adequate digestive enzymes from the pancreas, which are instigated by acidic foods leaving the stomach. Fat digestion will be compromised, too, because it depends on the acid's influence on the pancreas to secrete lipase and the gallbladder to secrete bile. Poor digestion of these macronutrients means poor absorption of our basic energy sources.

That's not all. Hypochlorhydria prevents adequate absorption of certain essential vitamins and minerals such as zinc, manganese, and calcium. This means that even if you're eating healthy, cruciferous vegetables—such as broccoli, kale, and cauliflower—that are known for their powerful detoxification properties, you won't be getting the full array of their amazing benefits, as you don't have enough stomach acid to break them down.

Some people rave about eating foods rich with the "energy vitamin" B_{12}—an essential building block the body needs to thrive—but even that is useless to you, as it can't be liberated from your food without adequate stomach acid. It's been estimated that 10 to 30 percent of adults over the age of 50 have difficulty absorbing vitamin B_{12} from food.[3] Low stomach acid is often responsible for this worrisome condition.

To top it all off, low stomach acid can also leave you constipated. When you digest food, it's moved through your intestinal tract by rhythmic contractions—imagine a worm slinking across the ground. Poor stomach function impairs these contractions. The result? You can't "go" when you want to.

We gradually lose stomach acid as we age, and we make things worse thanks to the usual culprits—stress and bad diet. It's not something we can afford to sacrifice. Without it, we're nothing but stinky, sluggish messes.

The Digestion Disaster

I really feel sorry for our stomachs. From childhood to death, those of us in the so-called developed world spend hours each day feeding ourselves more or less whatever we feel like eating: cookies, cereals, chocolate, coffee, takeout food, Big Gulp sodas . . . the list goes on. Amazingly, we somehow expect our stomachs and digestive system to break it all down and package it in a tidy little bow for our bodies to absorb. It just doesn't work that way.

When you think about it, we humans are quite fragile. Abuse another human being and you scar that person for life. Abusing your digestive system creates the same problem. In all my years working with clients, I have hardly ever seen anybody who has adequate digestive vitality. A huge number of today's energy, weight-loss, allergy, and autoimmune problems begin with sluggish digestion. Considering how I ate for the first 20 years of my life, I'm even amazed that I'm alive today to share this information with you.

When I was really young, I would get the worst stomach cramps. They were so bad that I had to leave the table and run upstairs to lie on my bed with my legs elevated up against the wall, in the hopes that I could "fart" away the pain. For years I had no idea what was going on. Neither did my parents, nor did my family doctor. It wasn't until I started learning about how certain foods like wheat, sugar, and dairy can wreak havoc on the stomach that I put two and two together.

The problem was that my diet was a mess. I was eating boatloads of cereals packed with sugar and drowned in 2 percent milk. I could survive on grilled cheese sandwiches. It's no wonder I suffered from low energy, bad eczema, allergies, and eventually an autoimmune condition.

Those little stomachaches that we encounter as children are the first signs of something far worse to come—a broken-down digestive system. And this is the beginning of almost every disease.

If you've ever fed a baby and the food didn't agree with him or her, what happened next? The baby would naturally vomit or have diarrhea as a way of purging that food. Believe me, your body knows what it's doing. Being able to understand its messages is another challenge.

In adulthood, if your stomach were continually exposed to irritating foods, causing incessant vomiting and diarrhea, eventually it would *harden* as a way of protecting you from dying. Continue abusing it with the wrong foods and it still gets shocked, but it no longer goes into full-blown contractions to elicit vomiting or diarrhea. By this point, your stomach has accepted its fate and can do nothing more than receive the continued onslaught of irritating foods. But now, its messages to you have been muted—unless you listen really, really carefully.

It tries to get your attention by making you nauseated and giving you heartburn, indigestion, gas, bloating, and even ulcers. Sadly, instead of listening to these cries for help and supporting their stomachs, most people seek out drugs to silence the symptoms. Many of these drugs suppress the very acid your stomach needs in greater quantities.

I'll never forget one TV commercial in particular that features a man suffering with acid reflux because of the garbage foods he'd been eating. Instead of cleaning up his act, he turns to an over-the-counter antacid (the product promoted in the ad), which soothes his tummy, allowing him to eat all the chicken wings, fries, and desserts he pleases.

This message is far too pervasive in our culture. Thinking we can take a pill to mask unpleasant symptoms so that we can continue enjoying our favorite foods (in other words, abusing our health) is ludicrous. These quick fixes only mask the growing threats to our health that sap us of the vitality we need to live the lives we truly want. Pushing our bodies past their limits and eating anything we please only gets in the way of the fulfilling lives we're seeking.

■ ■

If there's anything you should take away from this journey into your innards, it's that any food or process that makes your digestive system work harder than it's supposed to is going to rob you of your energy. It's most likely the primary reason you feel tired all of the time. You have to fix your digestion if you're going to restore the energy you've lost.

We could explore digestion for hours, but there are other pieces to this puzzle, and it's time to put them all together. It's time to bring you back to life.

Take the Digestive Vitality Test

How do you know if your stomach is broken and screaming for help? What are the signs and symptoms of hypochlorhydria? The following test will give you some immediate answers.

Give a 0 if a symptom, sign, or habit does not apply, 1 for mild or rarely occurring, 2 for moderate or regularly occurring, or 3 for severe or frequent.

	Do you ever feel bloated after eating?
	Do you ever get tired after a meal, especially one that contains heavier protein?
	Do you get gassy or belch upon eating or afterward?
	Do you have bad breath?
	Do you have longitudinal striations on your fingernails?
	Do you ever have undigested food in your stool?
	Do you eat when rushed?
	Do you eat in front of your TV or computer?
	Do you eat when stressed?
/27	TOTAL

Results:

- Score between 0 and 9 = digestive strength appears okay.
- Score between 10 and 18 = start increasing your stomach's digestive vitality.
- Score between 19 and 27 = immediate action required to properly digest foods.

Part II

THE ENERGY RESURRECTION

Chapter Five

WHAT TO EAT FOR ALL-DAY ENERGY

Here, we get to the heart of things and really get to work. This is where you turn things around and start bringing your energy back. It all begins in your kitchen. Your food is truly your fuel, and I'm going to help you turbocharge your life by changing your outlook and approach to eating.

Now, despite what you may have read or heard elsewhere, I hope you realize that no single "energy" food will override the quality of your overall diet. You can eat the latest superfood as much as you like, but if the rest of your diet isn't healthy and alkaline, you won't feel energetic—plain and simple. We're going to be reformulating your entire eating plan so you can feel more energetic and improve your health. This isn't about quick, desperate fixes.

Throughout this book I've shown you that establishing a slightly alkaline environment within your blood is crucial to good health and unstoppable energy. What's also great is that when you begin eating more alkaline foods, you inevitably increase the nutrients coming into your body. Since we're removing the processed garbage from your diet and putting more focus on nutrient-rich plant foods, it's almost impossible not to get what your body is starving for.

The All-Day Energy Diet Food Spectrum presented in this chapter is specifically designed to drive more alkalinity and energy into your body! It proposes eating larger quantities of all vegetables and some fruit—most of them in their raw state—along with moderate amounts of healthy meats, eggs, raw nuts, and good fats. In limited amounts, legumes and non-glutenous grains are suggested, especially for athletes or those who train frequently. The final portion, accounting for about 10 percent of food intake, is the leisure foods. We'll look more closely at each one of these groups in just a moment.

This isn't a revolutionary fad diet, nor is it a complicated way to "hack" your way into good health. It's actually shockingly simple and sensible. The goal of this food spectrum is not to measure and count calories but to look at your daily meals and ask yourself, *Have I got things in balance here?*

If you have steak (which is acid forming and tougher to digest) for dinner, you'll be looking to balance that out with a plate full of leafy green veggies and/or a salad. Additionally, a digestive aid like water with lemon or apple cider vinegar before the meal can help. Steps as simple as these can make a tremendous difference in how you feel and look.

Really, it's all about being empowered with the right knowledge so you can make smarter decisions for you and your family. Some meals and days might be more acidic, while others will be more alkaline. The overarching goal is to do your best to be aware of where you're at and how you're feeling every step of the way. Essentially, it's about awareness. For me, taking in plenty of raw plant foods throughout the day in the form of juices, smoothies, and salads, and then having a slightly heavier cooked or uncooked dinner, works best. For others, a more substantial breakfast or lunch might be preferable. You just have to experiment and see what sticks.

Again, it's all about honoring your body and figuring out what works for you. Just remember to keep your diet as alkaline as possible in the process. The good news is that by following the All-Day Energy Diet Food Spectrum, you will easily accomplish that objective. Once we get through these foods, I'll share my 7-Day Energy Reset Plan with you. This is a specific seven-day meal plan that has helped tens of thousands of people double their energy. You don't have to follow it for the rest of your life, but once you've experienced the surge of vitality that comes at the end of your first seven days, I'm pretty sure you'll want to stick with it.

Before we jump into each food category, let me lay down a couple of basic ground rules—fundamental principles, if you will.

Principle #1: Healthy Alkaline Blood Comes from Eating More Alkaline Foods

In case I haven't gotten the message across yet, this is one of the most fundamental things you need to understand: your body must be in an alkaline state to produce the energy you need. If you're primarily acidic, you're opening the door to all kinds of trouble.

Most of the foods you eat contain a wide variety of nutrients and minerals; however, your body processes them in different ways. Certain foods contain more alkaline minerals (calcium, magnesium, and potassium) compared to their protein and phosphorus content. These are the foods that alkalize and purify your blood. These are the foods that energize you!

PRAL—or potential renal acid load—is the way scientists determine whether a food is acidic or alkaline once metabolized. It's a very simple equation:

PRAL = [(0.49 x g protein) + (0.037 x mg phosphorus)] - [(0.026 x mg magnesium) + (0.013 x mg calcium) + (0.021 x mg potassium)]

Okay, that's not so simple. So let me break it down a little more:

PRAL = [protein + phosphorus] - [magnesium + potassium + calcium]

As you can tell, any food that contains more protein and phosphorus compared to those three alkaline minerals will be acidic. A negative value means the food has an alkaline load, whereas a positive value indicates an acid load. Let's look at some examples:

Food	PRAL (per 100 g)
Beet Greens	-16.7
Spinach	-11.8
Kale	-8.3
Swiss Chard	-8.1
Bananas	-6.9
Sweet Potato	-5.6
Quinoa	-0.2
Almonds	2.2
Buckwheat	3.4
Millet	8.8
Cashews	8.9
Beef	9.5
Sunflower Seeds	12.1
Low-Fat Cheddar Cheese	26.4

The data in this table is based upon the USDA database (rev. 18): http://ndb.nal.usda.gov.

Looking at this table, do you notice any trends? There are only a few foods displayed here, but you can clearly see that vegetables and fruit are alkaline forming, while most grains, nuts, seeds, and animal products are acid forming. This makes sense if you simply look back at the equation for PRAL.

The Bone-Chilling Truth about Dairy

And now, a little bit of information that you probably won't be so happy to hear.

Dairy (specifically, cheese) is the most acid forming of all foods. This is due to the large amount of protein and phosphorus it contains. Even though it also contains a substantial amount of calcium, its calcium benefits are few and far between, because the phosphorus-laden and acidic nature of dairy impairs calcium absorption and utilization in the body.

Comparing cow's milk to human breast milk, we see that dairy has four times more protein and about 1,018 mg of calcium per 100 g, while human breast milk only has 33 mg of calcium per 100 grams. Seeing that, we'd automatically believe that milk is great for your bones, right? Isn't that what we've always been told? Hold on a second. . . .

In the body, the relationships between minerals are more important than the quantity of any single one. For instance, the ideal calcium-to-phosphorus ratio is 2.5 to 1. However, cow's milk has a calcium-to-phosphorus ratio of 1.27 to 1. Human breast milk is much better—2.35 to 1, so it has more calcium in relation to phosphorus. Human breast milk is the ideal for humans. Cow's milk has a lot more protein and a lot more calcium because it's needed to feed baby cows, not baby humans or adult humans. Cows grow to about one ton in one year. Therefore, they need far more protein and way more calcium than we do. Humans take forever to reach 100 pounds. We don't need anywhere near that amount of protein or calcium. Furthermore, since phosphorus directly inhibits calcium absorption in the digestive tract, getting too much from food (think dairy, meat, and fizzy drinks) is not helpful for good bone health.

That's right—dairy can actually be bad for your bone health.

Today, the average person consumes about 740 mg of calcium per day. Ironically, the calcium intake of our Paleolithic ancestors has been theorized to be between 1,500 and 2,000 mg per day.[1] Remember, they never drank milk; it wasn't even available at that point in time. These were hunters and gatherers—not farmers—who ate large amounts of vegetables, tubers, berries, and wild game. They were active and exposed to natural sunlight daily. Not surprisingly, it's these foods—not milk—that are the ingredients for strong and healthy bones.

Are you scratching your head yet? If this doesn't sound right to you, here are few more insights that should make you second-guess whether you really need to be relying on dairy for strong bones:

- In Greece the average milk consumption doubled from 1961 to 1977, and shot up even higher by 1985.[2] During the period between 1977 and 1992, age-adjusted rate of hip fractures (an indication of osteoporosis) almost doubled, too.[3]

- In Hong Kong in 1989, twice as many dairy products were consumed as in 1966, and osteoporosis incidence tripled in the same period. Now, Hong Kong's milk consumption level is almost equal to that of Europe.[4] The same goes for its incidence of osteoporosis.[5,6]

- Americans, along with Australians and New Zealanders, consume three times more milk than the Japanese, and hip-fracture incidence in Americans is two-and-a-half-fold higher. Interestingly, among the segments of the population in America that consume less milk, such as Mexican-Americans and black Americans, osteoporosis incidence is twofold lower than in white Americans. It has been established that this is not due to genetic differences.[7, 8, 9, 10]

On top of all of these overwhelming "coincidences," I haven't even mentioned the potential negatives associated with the casein and lactose in milk or the hormones and antibiotics that may also be present. If you want an alternative to cow's milk, then consider almond or hemp milk. These are super-easy to make and are loaded with terrific nutrients like vitamin E and well-balanced essential fatty acids.

If you're wondering where you're supposed to get your calcium without dairy, then here are your best food sources:

Food	Calcium (mg)
Sesame Seeds (1 cup)	1,404
Collard Greens (1 cup cooked)	357
Sardines (3 oz)	325
Spinach (1 cup cooked)	291
Turnip Greens (1 cup cooked)	249
Kale (1 cup cooked)	179

This information comes from the USDA: http://ndb.nal.usda.gov.

As a reference point, 1 cup (250 mL) of 2 percent milk yields about 293 mg of calcium.

Notice a similar trend again—green veggies are your friend! Not only do they alkalize your blood, but they also provide adequate levels of calcium, which can realistically be attained in a day of healthy eating—that is, of course, if you're eating the All-Day Energy Diet way.

Still not convinced? Think about this: What do cows eat? Ideally, they graze on grass all day long. If they're eating grass and manufacturing their calcium from that, doesn't that tell you something?

Principle #2: Aim for 75 Percent Raw

We're taking quite a few of our cues from our Paleolithic ancestors. If you really think about how they survived, you'll quickly realize they weren't eating roasted chicken or grilled steak every night. Sure, they would cook meat they had successfully hunted, but for the most part, their diet consisted of the vegetables, fruits, and nuts they would have gathered. Furthermore, they would have eaten them in their raw state.

As we've seen, alkalizing your blood is quickly accomplished by eating more vegetables and nonsweet fruit. Within that range of food, you'll find that you can and will greatly enhance your energy and your health if you can eat more of it raw.

I can tell you this from firsthand experience and because I've seen the same with tens of thousands of others. Adding more raw foods to your diet is a surefire and amazingly quick way to feel more alive than ever. Mind you, I'm not recommending that you become a 100 percent hardcore raw food–ist, because there are still other cooked foods that greatly benefit you as well. That said, I'd like you to consider adding more raw into your diet. Ideally, your diet would primarily consist of raw foods.

Why?

Let me give you four reasons. The first two should make sense to you, but the last two will likely challenge everything you've ever been told about food. That's a risky proposition, but hey, I'm here to make you reconsider the common wisdom about food and health.

1. Nutrient Value

First of all, eating a food in its raw state leaves all of its nutrients intact. (*Note:* Let me be very clear—I'm not talking about eating raw meat. I'm referring to eating vegetables, fruit, seeds, and nuts in the raw.) In most cases, heating a food reduces its nutrient value. You might not know this, but grilling, baking, and even steaming destroys many of the vitamins, minerals, and phytonutrients that are found in your food.

The table below compares the typical maximum nutrient losses for common food-processing methods. This table is included as a general guide only. Actual losses will depend on many different factors, including type of food, cooking time, and temperature.

Typical Maximum Nutrient Losses (as compared to raw food)

Vitamins	Freeze	Dry	Cook	Cook and Drain	Reheat
Vitamin A	5%	50%	25%	35%	10%
Retinol Activity Equivalent	5%	50%	25%	35%	10%
Alpha-carotene	5%	50%	25%	35%	10%
Beta-carotene	5%	50%	25%	35%	10%
Beta-cryptoxanthin	5%	50%	25%	35%	10%
Lycopene	5%	50%	25%	35%	10%
Lutein	5%	50%	25%	35%	10%
Vitamin C	30%	80%	50%	75%	50%
Thiamin	5%	30%	55%	70%	40%
Riboflavin	0%	10%	25%	45%	5%
Niacin	0%	10%	40%	55%	5%
Vitamin B_6	0%	10%	50%	65%	45%
Folate	5%	50%	70%	75%	30%
Food Folate	5%	50%	70%	75%	30%
Folic Acid	5%	50%	70%	75%	30%
Vitamin B_{12}	0%	0%	45%	50%	45%

Minerals	Freeze	Dry	Cook	Cook and Drain	Reheat
Calcium	5%	0%	20%	25%	0%
Iron	0%	0%	35%	40%	0%
Magnesium	0%	0%	25%	40%	0%
Phosphorus	0%	0%	25%	35%	0%
Potassium	10%	0%	30%	70%	0%
Sodium	0%	0%	25%	55%	0%
Zinc	0%	0%	25%	25%	0%
Copper	10%	0%	40%	45%	0%

Courtesy of http://www.nal.usda.gov/fnic/foodcomp/Data/retn5/retn5_tbl.pdf.

Although I'm suggesting that you eat raw whenever you can, in some cases it may be more practical to lightly steam your food, especially if you can't stomach its raw version. Take cruciferous vegetables like broccoli and Brussels sprouts as an example. You can certainly eat them raw, but they'll be much easier to chew and digest if they are lightly softened first. They'll probably be a lot more palatable to you as well. That said, you can still enjoy their raw health benefits by blending them into raw soups, for instance. Remember, this isn't dogma. I'm encouraging you to think raw when it makes sense.

The easiest way I know of to add more raw foods into your diet is by doing any or all of the following three things: juicing, making smoothies, and preparing more salads. If you just did one of these things each day, your health, body, and energy would be utterly transformed. I'll show you how to make these easy to implement in the 7-Day Energy Reset Plan that follows.

You can have cooked foods whenever you want, but just remember: if you want the full power of nature's plant foods, then eat more of them in their raw form.

2. Hydration

The second reason to eat more raw plant foods is that they are loaded with water. This is important to you because you're probably walking around dehydrated—without even knowing it. In fact, many common symptoms that slow you down, like headaches and low energy, can often be remedied just by drinking more water. For some reason, though, many people don't enjoy the taste of pure water. By eating more fresh food, you don't have to worry about chugging water all day long. Simply make a fresh juice, smoothie, or salad, or snack on your favorite veggies and fruit, and you'll be sure to get more water in an "under-the-radar" fashion.

Unfortunately, heating your food through cooking dehydrates it. Think of what happens to a big fat burger patty as it simmers on the barbecue grill. Initially, it's all plump and juicy, but it quickly shrinks in size as most of its water content is evaporated. Make sense? Pretty much all processed food is devoid of water. If you eat lots of foods out of boxes and packages, then I bet your body is thirsting for water.

The Color of Your Pee, and Why Salt Could Be Your Best Friend

The easiest way to determine whether or not you're dehydrated is by looking at your urine. If it is any color other than clear, then you know you're in trouble. The darker it is, the more dehydrated you are. This can easily be fixed by downing a big glass of water or eating more fresh food throughout the day. *Capiche?*

IMPORTANT: If you know or think you have adrenal fatigue (that is, you feel exhausted, urinate frequently, have low blood pressure, feel dizzy upon standing, have salt cravings, and so on), then one of the best things you can do is add a pinch of sea salt to your water. Sadly, salt has been given a bad rap from the heart-disease world.

Here's the thing—if you limit your intake of processed foods, you won't be consuming much sodium (salt). However, if your adrenal glands are compromised, then they won't be secreting enough aldosterone—the hormone that regulates your body's sodium and fluid balance. As a result, your body won't be able to hold on to the sodium that it needs to function properly. Every time you go pee, you're essentially flushing away sodium. And that's one of the reasons why you might have low blood pressure and feel dizzy upon standing: there simply isn't enough sodium to maintain adequate water volume in your blood.

For this reason, adding some healthy sea salt to your water and food can be one of the easiest things you can do to start feeling better. It will instantly allow your body to hold on to the water it needs to function properly. Simply use this easy-to-remember formula—for every 2 cups (500 mL) of water you drink, add ¼ teaspoon (a pinch) of sea salt. When you eat a wholesome, fresh-food meal, do the same. This is especially important if you're eating mostly raw plant foods, like I do, which are naturally low in sodium and high in potassium.

Now that I've discussed the irrefutable reasons for eating more fresh foods, it's time to get controversial. The following two reasons are somewhat controversial only because they are not readily supported by much scientific literature—yet. Does that mean that they aren't valid? No, not at all.

What's the difference between science and magic? Magic is something we simply cannot explain empirically. What's magic today could become scientifically proven tomorrow. Being a science nerd myself, I find it frustrating that so many people are closed-minded to anything that hasn't yet been proven by "research." Yet, the irony is that an increasing percentage of scientific research nowadays is completely biased due to financial interests and professors looking to secure tenure or future research grants. This isn't to say that all scientific research has been tainted, but much of it has. As such, I simply ask that you open your mind to the possibilities that exist outside of what's reported in the latest research papers.

What I'm about to share with you could in fact be the two most powerful reasons for eating more fresh food. After all, what you cannot see is often more powerful than what you can see. (Gravity, anyone?)

3. Energetic Properties

In 1932, Dr. Harold Saxton Burr, professor of anatomy at Yale University School of Medicine, demonstrated that plants, animals, and humans possess electromagnetic fields, or energy fields. Through the use of sophisticated equipment, these fields were shown to represent the vitality of the living organism. With Burr's methods, the health of organisms can be determined by the radiance of their constantly fluctuating energy fields.

Living organisms—including *you*—are more than their physical components. The "mind" isn't physical, yet it is arguably what makes us humans so incredible. This intangible life force or "energy field" is difficult for some to accept, simply because we cannot see it with the naked eye. However, practitioners of Reiki and chakra and energy therapists can detect and work with these invisible fields like a builder with bricks. Just because we cannot see them doesn't mean they don't exist, right?

To understand the concept of life force—or *chi,* as it was called in the ancient East—you must remember that everything is ultimately energy. Everything in this world and everything in this universe is connected through an infinite web of energy. Trees, houses, cars, animals, food, cement, and humans—all are energy. The only difference is that your energy vibrates at a different frequency than that of other living organisms and objects.

You are an electromagnetic being. If you don't believe that, try sticking your finger into an electrical socket. Actually, please don't take me seriously—you'll get shocked! This simple fact shows you that your body conducts energy. If it didn't, you'd be able to stick your finger into that socket with no worries at all. Another example is acupuncture, which helps the healing process by opening up blocked energy meridians within the body.

Keeping this in mind, is it possible that you can conduct energy from food? I certainly believe so. Intrinsically, I think you do, too.

Unlike many dietitians and many other health advocates, I don't count calories. In fact, I don't recommend you do either. Here's why—counting calories usually means that you're eating foods that you probably shouldn't be eating too much of in the first place. And unlike so many other "gurus," I will never lie to you and tell you that you can eat whatever you want if you hope to look and feel better. You simply can't! Don't expect to feel more energetic and lose weight if you're eating creamy pastas, chocolate cake, ice cream, and other garbage foods—even if you're eating less of them. If you want to feel alive, you need to eat fresh food that is naturally vibrating at a higher energy, period.

Thankfully, a friend and colleague of mine, Christopher Wodtke, has been able to capture the energy field emanating from various foods through his Kirlian photographs. Looking at these images is the closest we can get to actually *seeing* the energy of the foods we eat. The images are truly mind-blowing. You can see them at: www.alldayenergydiet.com/kirlian.

What you'll notice is an *electricity* of energy emanating from the non-cooked foods that is much more muted in their cooked counterparts. It's quite spectacular.

Considering the prominent difference between raw- and cooked-food images (if you didn't check them, do so now), which do you imagine would energize you to a greater degree? I think the answer is pretty obvious.

4. Enzyme Content

Ask raw food–ists why eating raw is good for you and they'll almost always tell you that raw foods have higher amounts of food enzymes, which make digestion easier. Is that true? Well . . . partly.

It's true that cooking foods denatures their proteins, including enzymes (which are proteins). As a result, eating more raw could prevent those losses. But do the enzymes naturally found in fresh food really help digestion? Doesn't the body produce its own enzymes to do the job? The answers to these questions are still debatable, so let me present both sides of the story for you.

Enzymes are proteins that accelerate metabolic reactions in the body. They are involved in every single bodily process, from digestion and thinking to heart function. As I've mentioned, the three main categories of enzymes are digestive enzymes (ones that work on food digestion), metabolic enzymes (involved in every other bodily process), and food enzymes (enzymes inherent in fresh, raw foods).

Enzymes are key to life. Without them, we die, according to some experts. In his classic book *Enzyme Nutrition,* Dr. Edward Howell introduces the idea of an "enzyme bank account," which essentially means that we make enzyme deposits and withdrawals in our body based on the foods we eat.[11] The goal is to make as many deposits (by eating raw foods) as possible to offset any withdrawals from enzyme-void cooked foods. The underlying premise of this idea is that contrary to popular biological teachings, the human body has a finite number of enzymes. If we use them up, we die. To be honest, I don't know if this is true or false, but I've kept my mind open to the possibility.

Part of Dr. Howell's argument is that by eating more fresh, enzyme-rich foods, you alleviate the strain on your pancreas, which is responsible for producing the digestive enzymes in your body. In so doing, you conserve more of your body's precious enzymes, while the enzymes present in the foods you eat do most of the digestion.

Whether or not Dr. Howell's work is 100 percent true, the important thing to take away is that enzymes are vitally important to our digestion and overall health. Enzymes are essential biocatalysts, as they speed up the action of chemical reactions. Without them, the chemical reactions within us would be so slow as to make life as we know it impossible. The thousands of enzymes we require are either created as a result of our own enzyme production, which is arguably finite in nature, or attained from ingesting fresh, enzyme-rich food.

Enzymes are heat sensitive, which means they become denatured above 118 degrees Fahrenheit or so. Thus, cooked foods have virtually nothing in the way of enzymes. Furthermore, enzymes operate based on a "lock-and-key" system, meaning that they are very specific to the substrates they act upon. For instance, proteases break down protein, lipases break down fats, and amylases work on carbohydrates.

Yes, enzymes are important. And yes, you get more of them by eating raw foods. In the supplement section that follows, I'll also show you how supplemental digestive enzymes can not only improve digestion but also help repair many "broken" things inside your body. Enzymes are truly powerful and are just one more reason why eating more fresh, raw food is important to good health and all-day energy.

■ ■

In sum, adding more raw foods into your diet is beneficial for your energy levels and health for four main reasons:

1. They have a greater nutrient density so that you get more nutrients from the foods you eat, without all the calories.
2. They contain more water so that you stay well hydrated.
3. They vibrate at a higher energy level that resonates with your body so that you feel more energized.
4. They contain large amounts of enzymes, which are necessary for better digestion and metabolic repair throughout your body.

As I said earlier, I'm not preaching a 100 percent raw diet to you. Although some raw food–ists will disagree, no single diet (raw or other) will work for everyone all of the time. However, there are fundamental principles—like eating more raw plant foods—that will give you (and pretty much anyone) more energy and improve your health without fail. Basically, my message to you is this: eat more raw foods so that you can benefit from the four energy-boosting qualities that we just covered.

We've now covered two important principles to keep in mind when choosing your foods: eat more alkaline foods, and add more raw foods to your diet. Now, let's move on to the All-Day Energy Diet Food Spectrum and its various food groups.

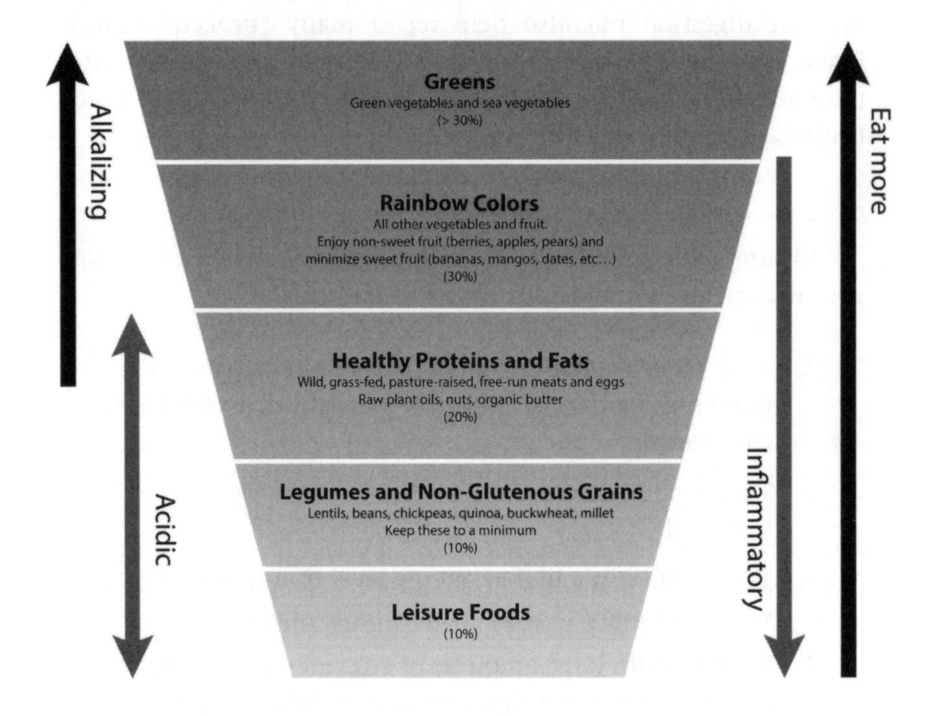

Greens

There's a good reason why your parents always told you to eat your greens when you were little. Even if you protested then, and still don't like them now, there's a simple fact you can't argue with: greens are the most important group of foods available on our planet.

When I speak of greens, I mean all plant-based foods that are green in color, and that goes for both land- and sea-based plant life. The green group occupies the top part of the Food Spectrum because they are the most alkaline of all foods and contain the highest nutrient density. They are also the richest source of chlorophyll, the pigment that gives plants their green color and captures energy from the sun. Chlorophyll is to plants what oxygen-carrying hemoglobin is to us: it assists in producing energy.

Chlorophyll is the lifeblood of plants and only differs from hemoglobin in that it is bound to magnesium instead of iron. Since these two molecules are so similar, consuming chlorophyll has been noted to provide blood-purifying and energizing properties. While the process isn't quite as simple as substituting the magnesium molecule in chlorophyll with an iron molecule to turn it into hemoglobin, there is evidence of the blood-building characteristics of chlorophyll-rich foods.

As far back as 1926, research suggested a relationship between the chlorophyll component pheophytin and hemoglobin generation.[12] Later studies indicated that feeding chlorophyll-rich foods to rats triggered the regeneration of red blood cells.[13] Yet another early study was conducted in which 15 patients with iron-deficiency anemia were fed different amounts of chlorophyll along with iron. It found that when chlorophyll and iron were administered together, the patients' number of red blood cells and their level of blood hemoglobin increased faster than when they were given iron alone.[14]

Chlorophyll helps our red blood cells (RBCs), and these RBCs carry oxygen to the cells throughout our body. It seems to make sense that consuming chlorophyll (via green foods or supplementally) can help our energy levels, at least indirectly.

Green leafy vegetables are also the best source of the highly alkalizing minerals calcium, magnesium, and potassium. These minerals help neutralize the acid-forming compounds in certain foods, which keeps our blood within its optimal, alkaline pH of 7.4. In turn, this goes a long way toward keeping us as healthy as possible.

A large review of the literature in 2004 looked at the relationship between fruit and vegetable intake and the incidence of cardiovascular disease, cancer, and deaths from other causes. They included intake from 71,910 females and 37,725 males. Of the food groups investigated, green leafy vegetables had the strongest protection against major chronic disease and cardiovascular disease. Essentially, the more greens consumed, the lower the risk of all disease and overall death.[15]

Pretty awesome, isn't it? Your parents were right.

The greens group includes, but is not limited to, the following vegetables. Ideally, they should all be eaten organic if possible:

- *Leafy greens*: kale, Swiss chard, collard greens, beet greens, mustard greens, spinach, bok choy, watercress, mache, lettuce, arugula
- *Denser green veggies*: broccoli, Brussels sprouts, peas, green beans
- *Grasses*: barley grass, wheatgrass
- *Herbs*: parsley, basil, cilantro
- *Sea vegetables*: dulse, nori, kelp, arame, Irish moss, hijiki, kombu
- *Blue-green algae*: spirulina, chlorella

Even though green veggies are jam-packed with tons of incredible nutrients, they are not a significant source of calories. That means that you can eat them all day long without worrying about packing on fat. It also means that surviving on greens alone is not a long-term strategy. Drinking nothing but green juice for an entire day can be terrific for your energy and overall health, but it's certainly not something I would advise doing 365 days per year; you'd soon find yourself withering away to nothing. That's why you'll want to include foods from the following groups to bring in the bulk of your calories and a different spectrum of nutrients.

Rainbow Colors

This second group contains all other vegetables and fruit. Little explanation is needed to describe the plethora of health-promoting benefits that come from eating vegetables and fruit of varying colors. A good rule of thumb is to combine lots of different colors in your meals, as this will ensure a wide variety of phytonutrients, which help prevent and fight various diseases. Need proof?

In 1996, a massive review of the scientific literature, including 206 human and 22 animal studies, investigated the relationship between vegetable and fruit consumption and risk of cancer. The results were conclusive: a protective effect of greater vegetable and fruit consumption was consistent for cancers of the stomach, esophagus, lung, oral cavity and pharynx, endometrium, pancreas, and colon. And the types of vegetables or fruit that most often appear to be protective against cancer were raw and cooked vegetables—specifically allium vegetables (onions and garlic), carrots, green vegetables, and cruciferous vegetables—and tomatoes.[16]

Eating more vegetables and fruit will also ensure that you meet your daily recommended intake of 35 g of fiber. Fiber is so important for several reasons. It helps to improve cardiovascular health, improves intestinal motility and bowel function, helps you lose weight, stabilizes blood sugar, and gives you a sense of fullness. This is an abbreviated list, so just remember that fiber is critical to your health. Thankfully, you don't have to rely on formulated fiber powders or whole grains for your fiber. You will get plenty simply by eating eight to ten servings of vegetables and fruit each day.

Notice how I've said "vegetables and fruit," and not the reverse? That's because you really should focus on the vegetable side of the equation. For the most part, fruit is okay, but you can also go overboard with the wrong fruit, and thus the sugar you consume. To make things simple, focus on eating nonsweet fruit like berries, apples, and pears. Berries are loaded with incredible amounts of antioxidants and have a very low glycemic index, so you can eat them all day long without negative effects. Pears and apples are low glycemic index as well, but they contain higher amounts of fructose, which shouldn't be a problem unless you're drinking gallons of apple or pear juice on a daily basis (I'll explain this in more detail in a moment).

Unless you're a very active person (working out several times per week or involved in a high-level sport), do your best to minimize sweet fruit like bananas and other tropical fruit. Eating too many of these fruits can wreak havoc on your blood sugar and aggravate gut-flora imbalances like candida. As a reminder, when blood-sugar levels rise, so too does insulin. Since insulin is a storage hormone, it will store all that excess sugar in your fat cells. With rising and falling blood sugar and insulin levels, you'll also fall into the energy-roller-coaster trap again.

The rainbow colors group includes (but is not limited to):

- *Colored vegetables:* carrots, beets, sweet potatoes, cauliflower, bell peppers, zucchini, eggplant, squash, onions, garlic, ginger
- *Nonsweet fruit* (as much as you'd like): berries (all), apples, pears, tomatoes, cucumbers
- *Sweet fruit* (limited amounts): bananas, dates, figs, papaya, mangoes, pineapple, watermelon, mangosteens, oranges, grapefruit, kiwis, cantaloupe, honeydew melon, grapes

Before going on to the next food groups within the All-Day Energy Diet Food Spectrum, I want to bring up two more discussion points:

- **First,** we'll quickly glance at which plant foods are best to get organic.
- **Second,** we'll look at what I believe to be the most powerful way to improve your health, lose weight, and boost your energy: drinking more freshly made smoothies and juices. Since we just discussed the

importance of eating more vegetables and some fruit, you should know that the easiest way to get more of these important plant foods into your diet, especially if you don't each much of them now, is through juicing and making smoothies.

Organic—When Does It Really Matter?

We all know that toxins and chemicals are lurking in the food we eat and the water we drink. It's no surprise, then, that our bodies have become increasingly toxic and prone to disease. We have reached our toxic threshold.

The allure of organic foods is that they are supposed to be free of pesticides and chemicals, which none of us want in our bodies. Of course we should absolutely consume more of them for this very reason, but this isn't realistic for many people, as organic produce tends to be more expensive and not as readily available as conventional produce. As such, it certainly helps to know that not all produce needs to be of the organic variety.

If that's the case, are organics really that much healthier than conventionally grown foods? Are they more nutrient dense? Well, the consensus in the literature is that the nutrient differences are small. Nonetheless, it's helpful to know the following two findings, which are pretty consistent:

1. The majority of studies on this topic have shown that organic crops appear to be higher in vitamin C and some essential minerals (calcium, magnesium, iron, and chromium).
2. Phytonutrients (lycopene in tomatoes, polyphenols in potatoes, flavonols in apples, and resveratrol in red wine), those important nutrients that fend off disease, tend to be higher in organic foods.

These two findings make sense because they indicate that in order to survive, plants are required to produce more antioxidant and disease-fighting nutrients in the absence of pesticides, herbicides, and other man-made protective agents. The good news is that we benefit from these additional important nutrients as we ingest these plant foods in their organic state.

In a perfect world, we'd all be eating and drinking organic produce all the time. If you can't, you need to know what conventional produce is safe to eat. My good friends Mira and Jayson Calton make that decision process easy for you in their books *Naked Calories* and *Rich Food, Poor Food,* in which they list produce items you should buy organic at all costs. They call them the "Terrible 20." They also provide a list of the "Fab 14": conventionally grown produce that is okay for you to eat, even in its nonorganic state.

The Terrible 20—Buy These Organic

Peaches	Lettuce
Apples	Cucumber
Sweet Bell Peppers	Green Beans
Celery	Hot Peppers
Nectarines	Potatoes
Strawberries	Kale/Collard Greens
Cherries	Hawaiian Papaya (potential GMO)
Blueberries	Zucchini (potential GMO)
Grapes	Yellow Squash (potential GMO)
Spinach	Corn (potential GMO)

Note: GMO refers to genetically modified organism—an organism/food whose genetic material has been altered using genetic engineering techniques.

Notice how most of these 20 foods would be eaten skin and all. Although there are a few exceptions, do your best to choose organic whenever eating foods in their entirety.

Since the following 14 foods are the least problematic from a pesticide perspective, don't stress about getting them organic if you don't want to.

The Fab 14—Okay If Nonorganic

Onions	Eggplant
Pineapple	Kiwis
Avocados	Cantaloupe
Cabbage	Sweet Potatoes
Sweet Peas	Grapefruit
Asparagus	Watermelon
Mangoes	Mushrooms

Now that you know which produce is best to buy organic, let's move on to the topics of juicing and making smoothies—your golden ticket to amazing energy and health.

Juices and Smoothies—What's the Big Deal?

Poor digestion, intestinal problems, and an average intake of 3.4 vegetables per day (for Americans)[17] means one thing—a whole lot of people are experiencing very little and inefficient nutrient absorption when they eat. Many of these problems can be corrected by the addition of more liquid to your diet. Think about it: you make one green juice and—bam!—you've just ingested six to eight servings of veggies. Try doing that otherwise. It's doable, but definitely more time-consuming and tougher to extract all of those amazing nutrients.

This is so powerful that I challenge you to make at least one green juice or smoothie each day. It's the easiest and best thing you can do for your health, energy, and waistline.

Will you take me up on the challenge?

The main difference between a fresh-pressed juice (not one that is bottled and bought in a store) and a smoothie is that juices contain none or very little of the fiber of the fruit or vegetable that was juiced. On the other hand, smoothies are liquid versions of whole food; all of the important fiber is intact.

The benefit to having no or little fiber is that the nutrients from the juice are generally more readily absorbed by our bodies. Fiber (aka cellulose) is the plant's "cell wall," which our human digestive tract cannot break down because we lack the enzyme cellulase. As amazing as plant foods are for our health, the catch-22 is that sometimes we can't completely absorb their full spectrum of nutrients because of that pesky fibrous wall. That's where juicing comes in handy. Juicing strips away the fiber and leaves us with pure liquid nutrition that can easily be absorbed into the body.

It's for this very reason that juicing is so beneficial for anyone with a compromised digestive and intestinal system (in other words, almost everyone). It's especially beneficial for those with IBS, Crohn's disease, and colitis—conditions that inhibit absorption of nutrients in the intestines.

On the other hand, smoothies are best for ensuring you get plenty of fiber and a minimal blood-sugar response from your liquid meal. Is one better than the other? I don't think so. They each serve a purpose, and I personally incorporate both smoothies and juices into my daily routine. Usually, I start the day with a green juice and then have a more substantial smoothie a little bit later on.

Note that making a fresh-pressed juice requires a juicer, whereas making a smoothie requires a blender. For most people, smoothies are easier and more convenient because blenders tend to be a household item and blending generally takes less time than juicing. Nonetheless, I recommend having a blender *and* a juicer so that you can enjoy the benefits of both.

The Most Important Thing to Remember about Juicing

Fresh-pressed juices should largely be composed of *vegetables*—not fruit.

Why?

Well, I've already mentioned fructose, right? Fructose is a type of sugar that is commonly found in fruit along with glucose. The problem with fructose is that it must enter a different and more complex metabolism process in your liver before your body can convert it into glucose.

Glucose, on the other hand, is much easier for the body to utilize, and it's the ultimate fuel source for your cells. The issue with fructose is that its metabolism (which occurs in the liver) is "rate-dependent," meaning that your liver can only handle so much fructose at once. Think about assembly-line workers who can't keep up with an increasingly fast conveyor belt. It's the same idea. Any amount above what the liver can handle starts yielding nasty by-products at an alarming rate. These include uric acid, lipid droplets (which can lead to fatty liver disease), and very low-density lipoproteins (VLDL). The latter become triglycerides in the bloodstream and can become very problematic for cardiovascular health.

Now that I've scared you away from eating fruit, I want you to rest assured that fruit is not the enemy. Fructose—in its processed form—is. Sodas, store-bought juices, and even fresh-pressed fruit juices are the biggest sources of fructose. Thus, these sources should be avoided or greatly minimized.

What . . . even fresh fruit juice isn't great?

Yup. Sure, fresh fruit juice contains lots of vitamins and minerals, but it's a liquid source of fructose that is not buffered by fiber, which means it will leave your digestive tract and enter your liver at a very fast rate, which means—yes, you got it—those unwanted triglycerides enter your bloodstream. Plus, whatever glucose is present in that juice will quickly spike your blood sugar, creating that undesirable blood-sugar roller coaster that we discussed earlier.

For the most part, I recommend adding *one* fruit (maybe two at the most) to your green juices to give them a bit of sweetness. The juices that are more heavily fruit based should only be consumed before or after rigorous exercise. The reason for this is that exercise speeds up your liver's ability to process fructose without shooting out all those unwanted by-products.

So don't forget: juicing is meant for vegetables, with the exception of adding more fruit before or after a workout, where you might need a little more sugar for performance purposes.

Try These Three Amazing Juices and Smoothies

SIMPLE GREENS (*JUICE*)

Makes 1–2 servings

1 pear
6 to 8 kale or Swiss chard leaves
½ lemon
1-inch piece of ginger

Run all ingredients through a juicer, serve, and enjoy!

CREAMY, ZINGY, LIMY SMOOTHIE (*SMOOTHIE*)

Makes 2 servings

1-inch piece of ginger
1 avocado, pitted and diced
1 apple, cored and quartered
1 cup/250 mL coconut water OR almond milk
Juice of 1 lime

Grate the ginger into a separate container and then squeeze the juice out of the grated bits into a blender.

Add all other ingredients to the blender, blend, and enjoy!

BERRY ALMIGHTY (*SMOOTHIE*)

Makes 3 servings

2 cups of your favorite berries
2 tablespoons almond butter
2 tablespoons hemp seeds
1 tablespoon flax oil
1 tablespoon maca (optional)
1 cup/250 mL almond milk

Add all ingredients to the blender, blend, and enjoy.

Clean Protein and Healthy Fat

Please allow me to clarify something for you—margarine is *not* a healthy fat. It is a man-made concoction that will poison you. You are better off eating organic butter any day of the week. In this section, I'll dispel some more myths about protein and fat and show you what to enjoy and what to avoid. This is perhaps one of the biggest areas of nutrition confusion, so I'm hoping what follows will give you some much-needed clarity and guidance for enjoying these foods in your pursuit of more energy and greater health.

Clean Protein

First, let me define what I mean by *clean protein*. For the most part, these are meats and eggs derived from animals that are labeled "wild," "grass-fed," "pasture-raised," "free-range," or "organic." For instance, if you eat salmon, choose wild, not farmed. When choosing beef, go with grass-fed instead of commercially raised. If choosing eggs, go with those from chickens that are free-range and organically fed, instead of those that have been cooped up and fed corn. Remember, you are what you eat, so if you're eating animal products, what they have been fed truly matters.

In other words, the animals you may choose to eat need to have had access to sunshine and a diet that comes naturally to them. These animals should not be caged or confined, and should not have been fed grains that have no place in their indigenous diet.

What's problematic is that many of the terms mentioned above aren't regulated. "Pastured" and "organic" are probably the two claims that I would trust the most, but all bets are off with the others. Unfortunately, it's quite common for "free-range" eggs to be anything but. When it comes down to it, the bottom line is this: knowing the farmers and how they raise their animals remains the best way to make sure you're getting the nutrition you're paying a premium for. Obviously, this is a utopian scenario that not everyone has access to, but you can certainly get close by buying more of your food from your local farmers' market.

With that said, I don't recommend eating copious amounts of animal products. As we saw earlier, they are acid forming and can be very hard on your digestive system because cooked protein is tough to digest. Picture what happens to an egg when you throw it in the frying pan—it goes from a gooey liquid substance (which is actually easier to digest) to a coagulated and solidified mass. Although the latter is more palatable, it's actually tougher for your stomach to break down cooked protein because there is less surface area exposed to the acid and enzymes in your stomach. Remember, anything that slows or impairs digestion will drain your energy. But again, I'm not recommending you start eating raw meat or eggs, okay?

The Palm Principle

If you do choose to eat meat, then please remember that you really only need as much as the size of the palm of your hand. Those massive steaks that you may have seen at a restaurant are going overboard; your body just can't handle them, especially if you're not producing enough hydrochloric acid. The good news with the palm principle is that it's relative to you. You might be a bigger or smaller person, and that will translate into the size of your palm. As a result, your portions will reflect that. For instance, a small, six-ounce sirloin steak will give you around 32 grams of protein. That's more than enough—assuming you can digest it properly.

What I am suggesting is that there are other great protein sources outside of the animal kingdom that can provide you with all the amino acids (building blocks of protein) that your body needs. The good news is that following the All-Day Energy Diet Food Spectrum can satisfy vegans, vegetarians, and carnivores. My hope is that the carnivores in the crowd will see the value in consuming slightly less animal protein and look at some of the following plant sources, which in many cases are just as good.

First, let's consider the foods that are highest in protein. This table summarizes all high-protein foods that are All-Day Energy Diet approved:

Food	Protein (per 100 g serving)
Spirulina, 1 cup	60 g
Beef	34 g
Pumpkin Seeds	33 g
Lean Meats (Chicken, Lamb)	30–33 g
Hemp Seeds	31 g
Lentils, Raw Sprouted	26 g (9 g if cooked)
Almonds	21 g
Chia Seeds	17 g
Walnuts	15 g
Egg, Whole	6 g

To give you some perspective, most of us only require about 0.8 g of protein per kg (2.2 pounds) of body weight a day. So, if you weigh 180 pounds (82 kg), you only need 66 g of protein every day. This can be easily achieved by drinking a smoothie that contains almond butter, almond milk, and hemp seeds along with some berries, and then perhaps having a small piece of steak for dinner.

As you can tell, you don't need as much protein as you've been led to believe. Furthermore, as important as protein is, you won't turn into dust if your intake is low on a particular day. Your body doesn't keep track of the days of the week. As long as you're getting enough protein on average, you'll be getting everything your body needs. Whether or not you want to supplement with a protein powder is up to you, but based on metabolic needs, all that money lavished on protein powders might be better spent on top-quality foods.

Protein for Vegans

As I just showed you, there are some really great sources of protein outside of the animal kingdom. And to be quite honest, if you incorporate the right foods, you can easily meet your protein requirements with plant foods alone. Including almonds, walnuts, hemp seeds, chia seeds, pumpkin seeds, and sprouted lentils in your diet is very doable for anyone—vegan or carnivore.

So let's say you are a vegan or at least want to minimize your intake of animal products; how much of these plant-based protein foods should you be eating? When it comes to healthy nuts like almonds and walnuts, aim for no more than half a handful of either. Remember, these nuts are also higher in calories, so we don't want to go too crazy.

Here's a little trick that will boost the nutrient and protein profile of your almonds: soak them overnight in a bowl of water. When you wake up in the morning, strain them, and give them a good rinse. The almonds will now be plumper and more easily utilized by your body. You can keep the remaining soaked almonds in the fridge for about 48 hours.

You don't need to do this with most other nuts, like walnuts, cashews, or pecans, since they have a proportion of fat and may end up a little soggy after an overnight soak. For your chia and hemp seeds, the easiest way to incorporate them is by adding them into your smoothies or by making some of the simple recipes that you'll get in the 7-Day Energy Reset coming up in the next chapter.

The other nice thing about getting more of your protein from plant sources is that they can be slightly easier on your digestive tract. Because cooked animal proteins are more difficult to digest due to denaturation and coagulation that occurs with heat, adding more raw plant proteins to your diet can be a good idea. Just remember that if you are omitting animal products, you'll probably need more plant foods to make up the difference in protein. By that I mean that for some people it can be easier to eat a small steak (which provides about 30 grams of protein) than it is to eat two to three tablespoons of hemp seeds—unless, of course, you've got some great recipes to work from.

Six Myths about Plant Protein

Protein is one of the biggest areas of confusion and misinformation out there. This chart summarizes six big protein myths.[18]

Myth	Reality
Plant proteins are incomplete.	Does not matter as long as "average" dietary protein intake is complete.
Plant proteins are not as good as animal proteins.	Quality depends on the source and amino acid makeup of the food (regardless of if animal or plant).
Protein combining is necessary if eating plants.	Complementary proteins do not need to be consumed at the same meal thanks to the body's amino acid pool.
Plant proteins are not well digested.	Depends on the source but can be improved by soaking/sprouting and supplementing with digestive enzymes and hydrochloric acid.
Plant proteins alone are not enough to get adequate protein intake.	Not true as long as you get all your amino acids (and nitrogen) from suitable plant foods.
Plant proteins are imbalanced, and this compromises their nutritional value.	As long as you consume enough calories from different plant foods, you'll be fine. Imbalances arise when supplementing with individual amino acids.

Healthy Fats

If you're scared of fats, I would like you to reconsider. The whole low-fat craze that began in the 1980s really has done nothing much to lower obesity or even heart disease—in fact, it's made it worse, because all that removed fat is usually replaced with some type of belly-fattening sugar.

In addition to the obvious trans fats, there are other fats and oils that should absolutely be avoided. To make your life easy, I'll list them here for you:

Processed vegetable oils:

- Canola oil
- Soy/soybean oil
- Corn oil
- Cottonseed oil
- Grapeseed oil
- Safflower oil
- Sunflower oil

Delicate oils (polyunsaturated fats) like these are very susceptible to degradation by light, oxygen, and heat. All of these oils are exposed to those three elements in the heavy processing and refinement that they are subjected to. By the time they hit store shelves, they are dangerously rancid. To boot, they are also very high in inflammatory omega-6 fatty acids, which are not good in high amounts in your body.

You'll almost always find these oils in packaged and processed foods and also in any items at a fast-food restaurant. Therefore, it's simple—if you stay away from fried foods, fast foods, and store-bought foods that come in boxes and packages, you'll be giving yourself a great head start.

In addition to the above-mentioned fats and oils, I would strongly discourage the use of margarine. As much as margarine companies attempt to persuade you that it's good for your heart and healthier than butter, the truth is that it is not. Think of margarine as a solidified form of these nasty vegetable oils we just looked at. It's man-made plastic. Believe me, you're much better off eating butter.

One of the largest studies ever done on fats and heart disease—the Framingham study—tracked participants for 20 years, and among other measures, compared butter and margarine consumption and their effects on heart health. The study revealed that consuming just five teaspoons of margarine (in other words, what you could spread on two pieces of bread) per day significantly increased the risk of coronary heart disease when compared to eating the same amount of butter.

Now, let's turn our attention to the healthy fats that you absolutely want to make an integral part of your diet. If you're scared of fats, don't be—fats make up the membranes of each of the trillions of cells in your body. If your cell membranes are healthy, you will be, too. Fats are essential precursors for the production of many hormones in your body. They are required to transport fat-soluble vitamins (A, D, E, and K) and provide protection and insulation for your vital organs. Without fats, you would die. In fact, more than 60 percent of the human brain is made of fat—the quality of that fat is up to you.

So . . . the *right* fats are critical to good health.

The Good Guys

There are three categories of fats that you need to know about. Each one provides unique benefits to you. They are *saturated fats, monounsaturated fats,* and *polyunsaturated fats.* Let's look at each one in more detail.

Saturated Fats

Saturated fats are very stiff, stable, and solid at room temperature. They are predominantly found in the fat cells of animals and in tropical oils. Butter and coconut oil are two terrific examples of yummy and healthy saturated fats. The irony is that because they're primarily composed of short- and medium-chain fatty acids, your body can easily use butter and coconut oil for energy production, instead of storing them in your fat tissue. By contrast, the saturated fat found in a piece of steak contains longer fatty acids, which are less easily burned as fuel and thus more readily stored as fat.

Since these fats are very stable, they are less prone to damage from heat, oxygen, and light, which means they are the best fats to use when cooking.

And don't worry, these fats will not skyrocket your cholesterol or increase your risk of heart disease. Most cardiovascular problems come about from eating too many refined carbohydrates, which increase VLDL and triglycerides in the blood. Cholesterol and healthy saturated fats are not to blame. In fact, a ten-year study of more than 52,000 subjects, which was published in the August 2011 issue of the *Journal of Evaluation in Clinical Practice,* showed that women with high levels of cholesterol were 30 percent less likely to die from heart disease, a heart attack, or a stroke, than women whose cholesterol was normal.[19] Numerous other studies have shown similar results.

Since I'm a big fan of coconut oil, I'd like to give you some more insight into why it's so beneficial for your health and energy. Coconut oil is composed predominantly of MCT (medium-chain triglycerides, which in turn are composed of medium-chain fatty acids). The MCT in coconut oil makes it different from all other fats and gives it its unique character and healing properties.

First, MCTs are easily digested, absorbed, and put to use nourishing your body. Because MCTs are smaller, they require less energy and fewer enzymes to break them down for digestion. Therefore, there is less strain on the pancreas and digestive system, which means less energy wasted. Upon digestion, MCTs' individual medium-chain fatty acids are easily absorbed and burned as fuel, much like a carbohydrate. Thus, eating coconut oil is like putting high-octane fuel into your car.

Did you know that your body's cells, especially those in your brain, prefer ketone bodies instead of glucose as a primary source of fuel? Ketones are created by the liver in the absence of carbohydrates and have been shown to burn cleaner and more efficiently than glucose. Coconut oil, loaded with its MCTs, produces more ketone bodies per unit of energy than normal dietary fats,[20] which means more clean-burning, high-yielding energy for you, and without the nasty side effects of raised blood-sugar and insulin levels. Pretty awesome, right?

Furthermore, two of the main medium-chain fatty acids (MCFAs) found in coconut oil, lauric acid and capric acid, are highly health promoting. These two compounds are big-time antimicrobial agents and are helpful in preventing infection and illness. Lauric acid has also been demonstrated to be the single most helpful fatty acid in boosting HDL cholesterol (the good kind).[21]

How to Choose the Healthiest Butter and Coconut Oil

When selecting butter, remember that how the cow was raised is very important. That said, choose organic butter whenever possible to ensure you're getting the cleanest and healthiest butter. It might be more expensive, but it's absolutely worth it.

Coconut oil is best in its "organic, extra-virgin" state. This means that it has the benefit of being certified organic, made from fresh coconuts, cold-pressed, unrefined, chemical-free, unbleached, undeodorized, and unhydrogenated, and it actually tastes like fresh coconut. This kind of coconut oil is so tasty and good for you that you can eat it by the tablespoon. That's what I do every single day!

Monounsaturated Fat

Monounsaturated fats are slightly less stable than saturated fats and are liquid at room temperature. They are found mostly in avocados, olives, olive oil, durians, cacao beans, almonds (and most nuts) and their oils, and to lesser extents, some red meat. Monounsaturated fats are less stable than saturated fats yet are more stable than polyunsaturated. However, as with the latter, these fats (especially in oil form) can go rancid with excess exposure to heat, light, and oxygen. Therefore, when choosing and storing monounsaturated oil like olive oil, be sure to follow these guidelines:

- Choose extra-virgin.
- Make sure it comes in a dark glass bottle.
- Store it in a cool, dark location like a cupboard or pantry.

The quality of olive oil production and pressing really makes a difference when it comes to health benefits. Recent studies have compared the anti-inflammatory benefits of extra-virgin olive oil (obtained from the first pressing of the oil) to the anti-inflammatory benefits of virgin olive oils (obtained from later pressings). It was shown that the extra-virgin olive oil was able to lower inflammatory markers in the blood where virgin olive oil was not able to do so.[22] Considering that inflammatory markers in the blood are big risk factors for heart disease and impede proper brain-to-adrenal-gland communication, I'm sure you can see the importance of going extra-virgin.

Foods containing monounsaturated fats have also been shown to reduce the "bad" low-density lipoprotein (LDL) cholesterol, while possibly increasing the "good" high-density lipoprotein (HDL) cholesterol.[23]

Monounsaturated fats, mainly in the form of olives and olive oil, make up a good portion of the diets of those living in the Mediterranean regions. It's no wonder this population is one of the healthiest in the world. In fact, people in Mediterranean countries consume more total fat than in Northern European countries, but most of the fat is in the form of monounsaturated fatty acids from olive oil and omega-3 fatty acids from fish, vegetables, and certain meats like lamb. Meanwhile, their consumption of saturated fat is minimal in comparison. For example, up to 40 percent of the dietary calories of those living in Crete come almost exclusively from olive oil, yet their incidence of heart disease and colon cancer are remarkably low.[24] I wouldn't recommend consuming that much olive oil, but I think you get the point—good fat (in moderation) is good for you.

Simple Tips for Making Monounsaturated Fats Your Friend

Here are some simple ways to enjoy the benefits of monounsaturated fats. Don't worry about counting calories or adding in one tablespoon too many.

- *Simple salad dressing:* Combine 3 tablespoons olive oil, 1 tablespoon apple cider vinegar or lemon juice, sea salt, and pepper.
- *Avocado by the spoonful*: Cut an avocado in half and remove the pit. Squeeze in some lemon juice and add some sea salt and pepper. Grab a spoon and dig in.
- *Veggie lube*: Drizzle extra-virgin olive oil over your favorite raw, sautéed, or steamed vegetables before serving. This can also enhance the absorption of fat-soluble vitamins in the vegetables.
- *Olive skewers*: As a fun appetizer, take a long toothpick or small skewer and insert various olives over its length. This can be a neat idea for a dinner party.

Polyunsaturated Fat

Polyunsaturated fats (PUFAs) are dominant in the substance and oil of most seeds and plants like walnuts, sunflower, flax, and hemp, as well as algae and cold-water fish. But remember, not all PUFAs are healthy. Those nasty, overprocessed, and highly inflammatory vegetable oils that we discussed earlier should be avoided as much as possible. Instead, focus on the good, anti-inflammatory omega-3s.

The reason that PUFAs can be so good or so wrong is because they are highly unstable. Their molecular chains have a number of double bonds, which means that there are more *weak links* where oxidation and damage may occur. That's why those heavily processed vegetable oils we discussed earlier are so problematic. Because of their highly volatile nature, PUFAs are the most susceptible to light, oxygen, and heat. For that reason, never use these oils for cooking. Likewise, these oils should be stored in dark bottles and kept in the refrigerator. Upon opening, they should be consumed almost immediately so that exposure to light and oxidative damage are minimized.

Within the PUFA category there are two main players—omega-3s (alpha-linolenic acid) and omega-6s (linoleic acid). You've probably heard of them before. Their chemical structures are slightly different, but their impact in the body is quite polarized. Omega-3s are the most beneficial for human health because they reduce inflammation in the body. From an energy perspective, that's very important since excessive inflammation will rob your energy and divert it to internal healing and repair.

Upon being metabolized, omega-3s yield EPA, DHA, and anti-inflammatory prostaglandins—all of which play hugely beneficial roles for our health. By contrast, omega-6s tend to create more inflammation in the body by yielding arachidonic acid and inflammatory prostaglandins.

The problem today is that the ratio of omega-6 to omega-3 inside the human body has been grossly shifted. Ideally, that ratio should be about 3 to 1, but since the processed Western diet relies so heavily on omega-6 vegetable oils (soy, corn, safflower, and so on), it's been said that the ratio is now closer to 20 to 1. It's no surprise, then, that we're suffering from so many degenerative inflammation-related diseases like heart disease, obesity, and cancer.[25]

Needless to say, your goal should be to minimize consumption of omega-6 fats and increase your consumption of omega-3 fats. I've already shown you the vegetable oils (omega-6s) to avoid, so here let's focus on omega-3s. The good news is that this is a short list of very accessible raw plant oils that you can find at pretty much any grocery or health-food store.

- *Hempseed oil:* Provides the ideal 3-to-1 ratio of omega-6 to omega-3 and is also a great source of the hard-to-come-by gamma linoleic acid (GLA), which may be helpful for skin conditions. For easiest use, add one to two tablespoons to your smoothie or salad dressing.
- *Flaxseed oil:* Composed of a whopping 58 percent omega-3 and very little omega-6, which can be favorable for helping reestablish a proper omega-6 to omega-3 ratio. For easiest use, add one to two tablespoons to your smoothie or salad dressing.
- *Chia (oil and seed):* Contains 30 percent omega-3, and its fiber-rich seeds are easiest to incorporate into smoothies and homemade puddings.

As great as these oils are, there is one small issue—the human body has a tough time converting them into the highly beneficial EPA and DHA. In fact, research has shown that the body converts about *one percent* of plant-based omega-3s into DHA.[26] And EPA is not much better. Part of the reason for this is that omega-3 metabolism is negatively impacted by the presence of high amounts of omega-6—since they both compete for the same enzyme (delta-6-desaturase).

Are you starting to see the problem here? You can consume as much omega-3 as you want, but unless you start cutting out those unhealthy vegetable oils (that are high in omega-6), you won't see much benefit.

With that said, there are a few other options that can bypass this conversion problem and give you the ultimate EPA and DHA that your body needs. These are cold-water fish and fish oil and algae and algae oil.

Cold-water fish (and their oils) are rich in DHA, as are various algae. Basically, anything that comes from the sea will have a decent amount of DHA. By contrast, vegetarian diets typically contain limited amounts of DHA, and vegan diets typically contain no DHA. It's been shown that strict vegetarians and vegans have substantially lower levels of DHA in their bodies, and short-term supplemental omega-3s in the form of plant oils (like flax and hemp) have been shown to increase EPA, but not DHA. Thankfully, supplemental preformed DHA, available in algae-derived oils or capsules, has been shown to increase DHA levels.[27]

DHA, even more so than EPA, is of ultimate benefit to our health. One of the main reasons is that DHA is a primary structural component of the human brain, cerebral cortex, skin, sperm, testicles, and retina. It comprises 40 percent of the PUFAs in the brain and 60 percent of the PUFAs in the retina of your eyes.

DHA deficiency has been associated with cognitive decline and depression.[28] Many other studies have shown that higher DHA intake confers benefits such as reducing the risk of Alzheimer's, reducing depression in Parkinson's patients, reducing inflammatory cytokines, and slowing the rate of telomere shortening—which is a DNA-level marker of aging.[29]

Take-home message—get more DHA. The best way to do that is by eating more cold-water fish and algae, or their oils. Sadly, most fish today come with higher levels of bioaccumulated toxins, like mercury. The good news is that high-quality fish and algae oils are usually free of these toxins. I'll discuss this in the next chapter.

If you decide to eat fish, please follow these guidelines:

- Consume two to three times per week max.
- Choose wild caught (not farmed).
- Small fish (sardines, for instance, or anchovies) are generally better (and cleaner) than big fish (salmon).

Most cooked salmon contains between 500 and 1,500 mg of DHA and between 300 and 1,000 mg of EPA per 100 grams. But does the cooking of the fish destroy the quality of those delicate fatty acids? That's up for debate, but it might be something to think about. Other good fish sources of DHA include tuna, bluefish, mackerel, anchovies, herring, sardines, and caviar.

If eating fish isn't your thing, then you can always choose a good-quality fish or algae oil supplement. I'll discuss those in greater detail in the next chapter. In the meantime, just remember that because inflammation is a key component in the disruption of so many vital processes, including hormonal communication and cellular function, anti-inflammatory omega-3 fats should be consumed much more frequently.

For optimal health, you need between 2 and 9 g of omega-3 per day, depending on your current state of well-being. The more banged up and inflamed you are, the more you'll need. Within that recommendation, 1 to 3 g per day should be in the form of EPA and DHA. For most people, those requirements might be tough to meet via food alone, and that's why fish and/or algae oil supplementation can be important. We'll look at that in the coming chapter.

Non-glutenous Grains and Legumes

Let me start this section by saying that I'm completely opposed to eating wheat and glutenous grains (rye, barley, oats, and the like). There really is no nutritional merit to bread, pastas, or cereals. If you need more fiber, you can get that by eating more vegetables and nonsweet fruit. With that said, there are nutritious non-glutenous grains that you can enjoy. But please do so sparingly, as the All-Day Energy Diet Food Spectrum implies.

The reason why quinoa, buckwheat, millet, and amaranth are less problematic is because they don't contain gluten and are less irritating to the digestive tract. They also provide a much greater nutrient profile and may contain up to 30 percent protein, as in the case of quinoa.

The most popular legumes include peas, beans, lentils, soybeans, and peanuts. Since soy is heavily genetically modified and overly processed, let's not even consider that to be a viable food option. The other legumes can be healthy in limited doses. After all, legumes are among the best protein sources in the plant kingdom. And since they're relatively cheap compared to meat, eating more of them may be an alternative to meat for some people. Legumes are also a great source of fiber and heart-healthy carbohydrates.

Although this food group is part of the All-Day Energy Diet Food Spectrum, I would recommend that you minimize your consumption of both grains (yes, even non-glutenous ones) and legumes to about 10 percent of total calories. You don't have to measure this, but just keep it in mind as a ballpark figure when planning your weekly meals. The reason for this is simple—grains and legumes contain elevated levels of anti-nutrients like phytic acid and lectins. Obviously, if you were to eat grains that contain gluten (which I strongly recommend you don't), then you could also add gluten to that list as well.

The problem with these anti-nutrients is that they are inherent compounds found in most plants and seeds (in differing quantities) that enable the plant to survive. When ingested, these anti-nutrients defend themselves—for instance, by attacking the delicate lining of your gut—so that they avoid being digested. The ultimate game plan for most seeds is to bypass digestion so they can be *pooped out* and replanted into the soil so that the plant can propagate. That's how seeds are programmed.

Normally, as food passes through the gut, it causes very minor damage to the lining of the gastrointestinal tract. Thankfully, our GI cells repair themselves quite rapidly. However, lectins can blunt this speedy reconstruction, meaning that our intestinal lining becomes less secure. When lectins affect the gut wall, this may also cause a broader and unfavorable immune-system response to deal with the problem.

The good news is that the harmful effects of lectins and phytic acid can be mitigated (although not completely erased) by using traditional methods of preparation, like sprouting, fermenting, and soaking and cooking. The only legumes I would recommend sprouting are lentils and alfalfa. Not only do sprouted lentils taste a hundred times better than cooked ones, they also boast three times the protein of their cooked counterpart.

For most other beans, your best bet is to simply boil them (if using dry beans) or grab a can of organic beans that you can easily add to your favorite soup or salad. Red kidney bean poisoning, for example, is usually caused by the ingestion of raw, soaked kidney beans, which contain phytohaemagglutinin—a lectin. Therefore, be sure to cook red kidney beans. Doing so reduces their lectin content from 20,000 to 70,000 lectin units to between 200 and 400 units. That's a significant difference.

How to Soak and Sprout

Sprouting seeds, grains, or beans decreases their lectin content. Generally, the longer the duration of sprouting, the more lectins are deactivated. Here's how to do it:

- **Step 1:** Soak beans and legumes overnight (adding baking soda to the soaking water can help further neutralize the lectins).
- **Step 2:** Drain and rinse.

- **Step 3:** Using a sprouting jar*, sprouting tray, or even a common strainer, place your soaked grain/legume/seed inside and make sure that there is enough ventilation. Tilt the jar at a 45-degree angle and place it in your dish rack so that water from your seeds can drain properly.
- **Step 4:** During this process, be sure to rinse your soon-to-be-sprouts once or twice a day to maintain adequate moisture.

*You can easily make your own sprout jars by taking a wide-mouth glass jar and covering it with some cheesecloth (secured by an elastic band).

Depending on the seed/grain/legume you are sprouting, different harvesting times will apply. Refer to the following chart for some guidance.

Food	Sprouting Time
Lentils	1-3 days
Alfalfa	4-6 days
Buckwheat	1-3 days
Quinoa	1-2 days
Peas	2-4 days

Once the seed/grain/legume has finished sprouting, you may store it in a covered bowl or Tupperware container in the refrigerator. Most sprouts will keep for several days, but it's best to quickly smell them before eating to ensure they haven't gone bad.

If one of your goals is to lose weight, then you'll definitely want to keep your intake of legumes and non-glutenous grains below 10 percent of your total calories. One of the reasons for this is that lectins can bind to leptin and insulin receptors, thereby increasing resistance to carbohydrates, causing weight gain, and impairing your ability to lose weight.

With all that said, you might be wondering why I've even included non-glutenous grains and legumes in the All-Day Energy

Diet Food Spectrum. The truth is that small amounts of lectins and phytic acid will not kill you. Like with most things, the danger is in the dose, unless we're talking about gluten, which is dangerous even in small amountsto those who are sensitive to it.

I don't feel there's anything wrong if you want to enjoy the occasional lentil soup, millet porridge, or quinoa salad. You don't have to, but at least you have the choice.

My beef is with the fact that the entire "eat more whole grains" movement has completely shattered our health. Most grains that you find in cereals and packaged foods are heavily processed and wreak havoc on your blood-sugar levels. It's no wonder we haven't won the war on fat—we've been fighting the wrong enemy. Fat—at least the good kind—is not the problem. The problem is the copious amounts of grains (and thus sugar) we continue to consume, thinking they are actually good for us.

Leisure Foods

With all of the commonplace foods I take issue with, you'd probably be surprised to discover that I consider myself a foodie. I truly am! That said, although I really love food, I also value my energy and health. And since I enjoy going out to dinner and indulging every now and then, I tend to be slightly more liberal in my recommendations. Yes, I believe gluten is the devil, but am I 100 percent gluten-free all the time? At home, yes; sometimes not if I'm at a restaurant. Removing gluten is definitely important, especially if you have celiac disease, but the idea I'd like to get across is that moderation is important.

I often tell people, "The best diet is the one you don't know you're on."

If you're constantly counting calories and worrying about every little thing that you put in your mouth, you're going to be pretty miserable. If you deprive yourself of some of the foods you enjoy most, it's only a matter of time before you rebel and binge on these foods.

Yes, it would be ideal if we all had the willpower to stay off certain foods all the time, but the truth is that that's not realistic for most of us—me included.

However, if you are somebody who after reading is hell-bent on removing gluten, caffeine, and sugar from your diet, then more power to you. You'll be much better off for doing so, especially if you get to a place where you never want those foods again.

I grew up on breads, cereals, sugar, and cheese. They are my kryptonite. When I eat them (which is very, very rarely), my energy plummets and my body yells at me. It's taken me almost three decades to get to a point where I don't really want these foods anymore because of how much I value *how I feel.*

That being said, I have found some terrific and healthy alternatives that have satisfied my sweet tooth and satiated my occasional yearning for bread. That's why the All-Day Energy Diet Food Spectrum provides a nice 10 percent cushion for such foods. I call them leisure foods. In the original version of this program, 20 percent of food intake was allotted to this category, but upon deeper investigation I realized that 20 percent is a heck of lot. In fact, 20 percent of the wrong food can sabotage all the good work you've done, so that's why 10 percent allotted to leisure foods is sufficient.

So what *are* leisure foods? They are whatever you want them to be.

Now be careful—I'm not saying that you should follow the All-Day Energy Diet food principles 90 percent of the time and then go binge à la Michael Phelps on pizza, hamburgers, ice cream, and fast food. That would be the worst thing you could ever do. Instead, what I'm suggesting is that you allot two to three meals per week (if you want) to please that inner child in a way that is rewarding, yet healthy. After all, there are tons of gluten-free, sugar-free recipes that taste amazing and don't ruin your health.

Here are a few ideas:

Instead of This . . .	Try This . . .
Ice Cream	Coconut-Milk Ice Cream
Cupcakes	Almond-Flour Cupcakes
Brownies	Zucchini-Based Brownies (amazing!)
Hamburger with Bun	Hamburger on a Gluten-Free Bun
Pizza	Pizza with Gluten-Free Crust
Coffee/Latte/Cappuccino	Decaf Coffee/Latte/Cappuccino
Regular Pasta	Rice Pasta
Pudding	Chia-Seed or Coconut/Cashew-Based Pudding

You might be wondering where on earth you can get these recipes for "alternative" food ideas. Don't worry—I'll share the ones I have with you in the recipe section coming up. Where no recipe is provided, thankfully there's Google, which will give you hundreds of possible options for healthy leisure-food ideas. Simply do a quick search for things like "gluten-free ____ [specify food]" or "dairy-free ____" or "sugar-free ____" and you'll come across many amazing recipes that will please your taste buds without sabotaging your health.

So yes, you can have your cake and eat it, too!

Good Sweeteners?

If you need to sweeten your desserts, tea, or anything else, here's a list of what I would recommend. The following natural sweeteners are certainly better than artificial sweeteners like aspartame and sucralose.

- *Xylitol:* Naturally derived from the fiber within plant foods, it contains almost no calories and, surprisingly, actually helps fight cavities.
- *Stevia:* Derived from a natural plant in the sunflower family, it is 300 times sweeter than sugar without the impact on blood sugar. I use it in some of my recipes in very small amounts.

- *Raw honey:* Honey has many beneficial health properties but can range in glycemic index from low to high depending on the variety.
- *Pure maple syrup:* This sap-derived sweetener is loaded with antioxidants but comes with a higher glycemic index.

The neat thing you'll start to experience as you eat cleaner following the All-Day Energy Diet principles is that you'll have fewer cravings for traditional cheat foods. And, if and when you do eat them, your body will usually feel worse. Since humans are highly motivated by the avoidance of pain, experiencing unpleasant symptoms after eating unhealthy cheat foods can be a blessing in disguise. Experience these negative symptoms often enough and you'll question why you even need to eat these foods, especially when there are healthier and tastier alternatives out there.

The leisure-foods group is geared toward giving you the choice and flexibility to enjoy some of the foods that you've typically enjoyed, while also offering you the possibility to completely eliminate them. Just remember, it will be difficult to move forward in your growth if you're still dependent on foods that hold you back. For maximum vitality, no more than 10 percent of your meals should come from leisure foods. However, the cleaner your leisure foods, the more blurring there will be between what is and is not a "cheat" food. That's great, because you can then get to a point where more of your meals not only energize you but also truly satisfy that inner child.

■ ■ ■ ■

Chapter Six

THE 7-DAY ENERGY RESET

Now that we've covered the foods, you're ready to rock and roll! Here's what we're going to do first. For the next seven days, we're going to reset your body. Consider it like rebooting your computer so that it runs more quickly and without any glitches. For the next seven days, simply follow this proven meal plan and you'll start feeling like a million bucks before the end of the week.

Mind you, this seven-day meal plan is not how you have to eat for the rest of your life. It's simply a more alkaline-focused quick start to reset your health and energy. Got it? Each of the seven days has meals listed for breakfast, lunch, and dinner, along with a snack idea that you can incorporate if you need to throughout the day. The recipes for these meals are provided in the pages that follow.

	DAY 1	DAY 2	DAY 3	DAY 4	DAY 5	DAY 6	DAY 7
Breakfast	Smoothie: The Green Machine	Juice: Simple Greens	Fiber Starter	Juice: Green Fiesta	Smoothie: The Incredible Hulk	Morning Millet	Smoothie: Creamy, Zingy, Limy Smoothie
Lunch	Nori Wraps	Blueberry Morning	The Whole Enchilada Salad	Sprout Salad	Garlic and Spinach Soup	Smoothie: Blue Chia	Quinoa Salad
Midday Boost (optional)	Energy Greens	Energy Greens	Energy Greens	Energy Greens	Energy Greens	Energy Greens	Energy Greens
Dinner	Zucchini Pasta with Marinara Sauce	Broccoli-Kale Soup	Curry Kale Salad	Salmon, Asparagus, and Dill Salad	Quinoa Veggie Bowl	Vegan Veggie Sushi	Chickpea and Kale Bowl
Snack (optional—to be enjoyed at any time of day)	Raw Chocolate	Smoothie: Tangy Coco Spinach	Guacamole	Hemp Balls	Baba Ghanouj	Juice: Sweetly Romaine	Strawberry-Avocado Salad

Note: Please see the Resources section at the back of this book for information on my Energy Greens.

7-Day Energy Reset Recipes

Day 1

THE GREEN MACHINE

Makes 2–3 servings

1 head kale, Swiss chard, or dark lettuce
1 banana
1 apple
1 pear
Juice of 1 lime
2 tablespoons hemp seeds

Add all ingredients to your blender along with 2 cups of water. Blend and enjoy.

NORI WRAPS

Makes 2 servings

4 sheets nori
1 avocado, thinly sliced
1 mango, thinly sliced
1 handful alfalfa or pea sprouts
¼ cucumber, thinly sliced

Lay out 1 sheet of nori, moisten with a sprinkle of water, and place within it a desired amount of each ingredient.

Roll up the nori (with ingredients inside) into a wrap and enjoy. Repeat with the remaining ingredients.

ZUCCHINI PASTA WITH MARINARA SAUCE

Makes 2 servings

1 zucchini, skin removed
2 cloves garlic, diced
1 cup sun-dried tomatoes, soaked
3 cups chopped tomatoes
2 dates, pitted and soaked
¼ red onion, diced
2 tablespoons olive oil
½ handful parsley, chopped
½ handful basil
⅓ cup olives, pitted (optional)
Pinch of sea salt or kelp/dulse flakes

Using a vegetable peeler or spiralizer, shave the zucchini down to paper-thin noodles.

In a food processor, pulse all other ingredients until smooth to create the marinara sauce.

Place the zucchini noodles on a plate, top with the marinara sauce, and serve.

RAW CHOCOLATE

Makes six to eight 2 × 2 squares

½ cup cacao nibs
½ cup cacao powder
¾ cup raw cashews
1 tablespoon coconut butter
¼ cup agave nectar or honey
1 teaspoon vanilla

Put everything in a food processor and blend until thick.

Remove the blend from the food processor and shape into squares.

Store the chocolate squares in the freezer before serving.

Day 2

SIMPLE GREENS

Makes 2 servings

1 pear
6 to 8 kale or Swiss chard leaves
½ lemon
1-inch piece of ginger

Run all ingredients through a juicer, serve, and enjoy!

BLUEBERRY MORNING

Makes 1 serving

½ cup fresh blueberries
2 tablespoons chopped walnuts
2 tablespoons hemp seeds
2 tablespoons shredded unsweetened coconut
1 tablespoon ground flaxseed
1 to 2 cups almond milk

Combine the first five ingredients in a bowl, pour almond milk over the top, and enjoy.

BROCCOLI-KALE SOUP

Makes 4 servings

4 cloves garlic, diced
1 onion, diced
1 carrot, diced
2 celery stalks, diced
2 heads broccoli
1 bunch kale, stemmed
4 cups of vegetable broth

In a large pot, sauté the garlic, onion, carrot, and celery until lightly browned.

Add the broccoli and kale.

Pour in 4 cups of boiling vegetable broth and cover. Simmer for 45 minutes or until the vegetables are soft. Puree all ingredients in a blender, then serve and enjoy!

TANGY COCO SPINACH

Makes 2–3 servings

2 big handfuls spinach
1 banana
½ cup strawberries
1 to 2 stalks celery
1-inch piece of ginger
Juice of ½ lime
1 young coconut (water and meat)

Grate ginger into a small bowl or container, then squeeze grated ginger so that its juice can easily be poured into your blender.

Add all ingredients to the blender. Add 1 cup water, blend, and enjoy.

Day 3

FIBER STARTER

Makes 1 serving

½ cup berries or chopped apple
2 tablespoons hemp seeds
2 tablespoons chia seeds
2 tablespoons sunflower seeds
1 tablespoon ground flax seeds
2 cups almond milk

Combine the first five ingredients in a bowl. Pour almond milk over the top. Let the bowl sit for 2 to 3 minutes before serving to allow the chia seeds to absorb the liquid and expand.

THE WHOLE ENCHILADA SALAD

Makes 2 servings

1 head lettuce or mixed greens
1 tomato, chopped
1 avocado, cubed
1 orange or yellow sweet pepper, chopped
¼ large cucumber, peeled and sliced
¼ cup olives, pitted
¼ sweet onion or shallot, sliced
1 handful sprouts (alfalfa, radish, etc.)
2 tablespoons olive oil
Juice of ½ lemon
Pinch of sea salt

Combine all ingredients in a salad bowl, toss lightly, and serve.

CURRY KALE SALAD

Makes 2 servings

1 bunch kale
3 tablespoons olive oil
Juice of ½ lemon
1½ teaspoons fresh curry powder
1 teaspoon cayenne powder
2 cloves fresh garlic, minced
1 teaspoon fresh grated ginger
1 to 2 dates, soaked

Roll up the kale and chop it into thin strips. Place it in a salad bowl.

For the dressing, blend all other ingredients in a high-speed blender or food processor. Add water to thin out the dressing, if needed.

Toss the kale salad with the dressing.

Serve with your favorite piece of fish or organic meat.

GUACAMOLE

Makes 2–3 servings

3 ripe avocados
2 tomatoes, diced fine
½ red onion, minced
½ cup cilantro, minced
1 to 2 cloves garlic, minced
Juice of 1 lime
Pinch of sea salt

Place all ingredients in a bowl. Mash and combine them with a fork until they reach your desired consistency.

Serve with your favorite veggies.

Day 4

GREEN FIESTA

Makes 3–4 servings

1 head kale or Swiss chard
1 large handful spinach
1 head romaine lettuce
1 cucumber
10 to 12 celery stalks
1 apple
1 lemon

Run all ingredients through a juicer, serve, and enjoy!

SPROUT SALAD

Makes 2 servings

3 cups sprouts (broccoli sprouts, alfalfa, lentils, etc.)
1 cup cherry tomatoes, halved
2 tablespoons olive oil
1 tablespoon apple cider vinegar
Splash of lemon juice
Pinch of sea salt
Fresh ground black pepper

Combine sprouts and tomatoes in a bowl.

Whisk all other ingredients together. Combine vinaigrette with sprout mixture and serve.

SALMON, ASPARAGUS, AND DILL SALAD

Makes 2 servings

2 small salmon fillets
8 asparagus spears, bottoms removed
1 handful salad greens
3 tablespoons chopped fresh dill
¼ cup green olives, pitted
1 tablespoon capers
Juice of ½ lemon
2 tablespoons olive oil
1 tablespoon coconut oil
Fresh ground black pepper
Pinch of sea salt

Preheat the oven to 450°F.

In a hot pan, melt the coconut oil and add the salmon fillets. Let the salmon sear for 2 to 3 minutes and then switch sides. Place the salmon in the oven for 4 to 5 minutes so that it is tender but not dry. Remove it, let it sit for a moment, and then cut it into bite-size portions and put it aside.

Steam the asparagus for 3 to 4 minutes or until the desired softness is achieved. Alternatively, you c an enjoy it raw.

Mix all ingredients together in a large bowl and serve.

HEMP BALLS

Makes 8–10 balls

1 cup soaked almonds
1 cup sesame seeds, milled
¼ cup honey
½ cup coconut oil
2 cups hemp seeds

Place the almonds in a food processor and grind them until fine.

Add the sesame seeds, honey, and coconut oil. Process until everything is combined.

Move the ingredients to a bowl and add the hemp seeds. Form the mixture into balls, and place them in a container in the refrigerator before serving.

Day 5

THE INCREDIBLE HULK

Makes 2 servings

1 banana
1 cup spinach
2 tablespoons hemp seeds
2 tablespoons hemp or flax oil
1 tablespoon ground flax seeds
1 tablespoon cacao powder (optional)
3 to 4 drops of stevia

Add all ingredients to your blender, blend, and enjoy.

GARLIC AND SPINACH SOUP

Makes 4 servings

1 tablespoon coconut oil
1 bulb garlic, cloves cut in half
1 bulb garlic, cloves minced
2 bunches green onions, green parts only, minced
4 cups vegetable stock
1 bag spinach, roughly chopped

¼ cup minced parsley

3 tablespoons Bragg Liquid Aminos (optional)

Fresh ground black pepper to taste

Put the coconut oil and halved garlic cloves in a medium to large soup pot.

Sauté the cloves until they brown slightly.

Add the rest of the ingredients and simmer for 15 to 20 minutes. Serve and enjoy.

QUINOA VEGGIE BOWL

Makes 2–3 servings

1 cup quinoa

1 sweet potato, peeled and sliced

1 zucchini, sliced

1 eggplant, sliced

½ teaspoon kelp seasoning (or sea salt)

3 to 4 tablespoons olive oil

1 tsp fresh ground black pepper

½ head Swiss chard, kale (stemmed), or spinach

1 avocado, peeled and sliced

Preheat the oven to 350°F.

In a medium saucepan, add the quinoa to 2 cups of water and bring it to a boil. Reduce the heat and let it simmer until the quinoa absorbs all of the water and is puffy and soft.

Place the slices of sweet potato, zucchini, and eggplant into a glass oven tray and sprinkle them with olive oil, salt, and pepper. Place them in the oven for 20 to 30 minutes.

Wash the Swiss chard, kale, or spinach. Place in a steamer for 15 to 20 minutes.

Place the quinoa in a bowl and top it with the greens, roasted veggies, and sliced avocado.

BABA GHANOUJ

Makes 3 servings

1 eggplant, skin removed
2 to 3 tablespoons coconut oil
Fresh ground black pepper
Pinch of sea salt
½ cup ground sesame seeds or tahini
Juice of 1 lemon
2 to 3 cloves of garlic, minced
1 to 2 tablespoons olive

Preheat the oven to 350°F. Slice the eggplant lengthwise into ¼-inch slices.

Season the eggplant slices with coconut oil, fresh ground black pepper, and sea salt.

Place the eggplant slices in an oven tray and set them in the oven for 15 to 20 minutes, or until the eggplant is golden and soft.

Let the eggplant slices stand and cool for a few minutes.

In a food processor, add the baked eggplant, ground sesame seeds, lemon juice, garlic, and olive oil and process until smooth and creamy.

Enjoy with your favorite veggies or as a spread.

Day 6

MORNING MILLET

Makes 2 servings

½ cup millet
1 cup berries
½ cup coconut milk
1 tablespoons flax oil
2 tablespoons maple syrup (optional)

Add the millet to 3 cups of water and bring it to a boil. Reduce the heat and let it simmer until the water is absorbed and the millet is soft.

Pour in the berries and allow their juices to mix with the millet.

Remove from the heat and let it sit for 1 to 2 minutes before pouring the coconut milk over the top.

Top with the flax oil and maple syrup (if desired), mix, and serve.

BLUE CHIA

Makes 2 servings

1 cup frozen or fresh blueberries
1 to 2 tablespoons chia seeds
2 cups almond milk
1 teaspoon honey or maple syrup (or 1 to 2 drops of liquid stevia)

Add all ingredients to your blender, blend, and enjoy.

VEGAN VEGGIE SUSHI

Makes 10–12 rolls

2 cups cauliflower
1 handful soaked cashews, walnuts, or pine nuts
4 tablespoons ground sesame seeds
1 tablespoon ginger, minced
Juice of ½ lemon
2 to 3 sheets nori
½ cup pea or alfalfa sprouts
¼ cucumber, julienned
½ red pepper, julienned
2 green onions, julienned
1 avocado, sliced
1 mango, sliced (optional)
3 to 4 tablespoons tamari

To make the "rice," throw the cauliflower, soaked nuts, sesame seeds, ginger, and lemon into a food processor and pulse until the mixture resembles rice.

Cover ½ of the nori sheet with the "rice" and ¼ of the sheet with sprouts.

Lay your julienned vegetables over the other ½ of the "rice" layer.

Tightly roll the sushi using your hands or a sushi mat, and use a bit of water to seal the nori in place.

Let it sit for a few minutes before cutting the roll into 5 or 6 pieces.

Use tamari (healthier than soy sauce) to dip your sushi.

SWEETLY ROMAINE

Makes 1–2 servings

1 small apple
1 head romaine lettuce
1 carrot

Run all ingredients through a juicer, serve, and enjoy!

Day 7

CREAMY, ZINGY, LIMY SMOOTHIE

Makes 2 servings

1-inch piece of ginger
1 avocado, pitted and diced
1 apple, cored and quartered
1 to 2 cups coconut water or almond milk
Juice of 1 lime

Grate the ginger into a separate container and then squeeze the juice out of the grated bits into a blender.

Add all other ingredients to the blender, blend, and enjoy!

QUINOA SALAD

Makes 4–6 servings

2 cups quinoa
¼ cup flax oil
2 tablespoons olive oil
1 tablespoon apple cider vinegar
¼ teaspoon sea salt

¼ cup raisins
½ cup kalamata olives, halved
½ cucumber, chopped
½ red pepper, diced
1 pint mini tomatoes, halved (or 2 medium)
¼ red onion, diced

Place the quinoa and 6 cups of water in a pot and cook per instructions on the package. Once the quinoa is cooked, set it aside to cool.

Combine the quinoa with the rest of the ingredients and serve.

CHICKPEA AND KALE BOWL

Makes 2 servings

½ cup chickpeas
Pinch of sea salt
1 tablespoon curry powder
4 to 5 stalks kale, stemmed
2 to 3 stalks Swiss chard
1 tablespoon sesame seeds
2 tablespoons chopped almonds
1 tablespoon olive oil
Juice of ½ lemon
Fresh ground black pepper

If using raw chickpeas, warm them in a pot with a little bit of water, a pinch of sea salt, and the curry powder. Allow the chickpeas to absorb the curry powder until they take on a yellow tint. If using canned chickpeas, then simply heat them lightly in a pot with a little water along with a pinch of sea salt and curry powder. At the same time, steam the kale and Swiss chard so they soften.

Put the chickpeas in a bowl and top them with the steamed kale and Swiss chard. Sprinkle them with the sesame seeds and almonds.

Drizzle the olive oil and fresh-squeezed lemon over the top, and season with fresh ground black pepper.

STRAWBERRY-AVOCADO SALAD

Makes 2 servings

1 avocado, sliced
6 strawberries, stemmed and sliced
Juice of ¼ lemon
1 tablespoon balsamic vinegar
Fresh ground black pepper
Pinch of sea salt

Place the avocado and strawberries in a bowl.

Add the remaining ingredients, mix gently, and serve.

▪ ▪ ▪ ▪

Chapter Seven

SAFE SUPPLEMENTS AND SIMPLE SUPERFOODS THAT GIVE YOU LASTING ENERGY

I have to admit that I don't really enjoy taking pills. Come to think of it, who does? As I stated in the last two chapters, eating whole foods should be your main focus as you work toward feeling more energetic. That said, there are times when specific supplements can help you along on this path. In this chapter, we'll take a look at three categories of supplements that you can consider incorporating into your diet. Remember, these are called supplements for a reason—none of them can or should replace the foods you're supposed to be eating.

These energy-boosting supplements fall into three categories: *blood support, digestive support,* and *adrenal and stress support.* Blood support supplements are recommended for everyone, as are digestive support supplements, especially for those who suffer with sluggish digestion or intestinal issues. Adrenal and stress support is suggested for those who have full-blown adrenal fatigue or who are less able to cope with stress.

Chances are, if you're reading this book, you'll benefit from all three categories. Mind you, they're not absolutely necessary, but they can be very helpful, especially in combination with the All-Day Energy Diet nutrition guidelines you'll be observing.

Blood Support

Greens Powders

Since one of the main goals of this program is to alkalize your blood, it's essential for you to add more alkalizing greens to your diet. That's quite a shift for some people, especially those for whom eating a salad is nothing less than torture. Thankfully, there's an easier way: greens powders. These are powdered mixes derived from greens such as wheatgrass, chlorella, alfalfa, and spirulina—among others—that can be added to water for a quick, tasty drink. They can also give your smoothies a powerful, alkalizing boost. Furthermore, they're great to have on trips away from home when you might not have access to a blender or juicer.

Go to any health-food store and even a few supermarkets and you'll find a variety of greens powder blends on the shelves ranging from $20 to $100. As with anything else, not all greens powders are created equal, so you have to be very discerning when making your purchase. Most cheaper greens products are made from whole grass that has been cut, dried, and powdered. This results in a bulkier powder that is cheaper and doesn't mix very well with water. Even I don't enjoy drinking these types of greens—they just don't taste good at all.

On the flip side, there are premium greens powders that use grass *juice* powder. This means the grass has been cut and then juiced with a cold press. This juice is then dried at a low temperature, powdered, and packaged. This careful process yields a much richer and finer green grass blend with a higher overall nutrient density. It's also 100 percent raw. These are the types of greens powders you want. The easiest way to find them is by carefully inspecting the labels of any greens powder you're considering—it should say something to the effect of *wheat grass juice powder, barley grass juice powder,* or similar wording.

The most noticeable benefit of these higher-quality powders is how easily they mix into water. Powders that haven't been produced from juice tend to clump up in water, with little bits of fiber floating around. Who wants to drink that? That's never the case with juice powders, and what you're left with is a delicious green drink that will please your taste buds and purify your blood. (To save you the hassle of trying to find a decent greens powder, you can learn more about my Energy Greens in the Resources section at the back of this book.)

The benefits of alfalfa and barley grass include immune support, antifungal properties, and blood-sugar stabilization. They're also highly alkalizing, help fight free radicals, and purify the blood and liver of chemicals and toxins.

Spirulina and chlorella are two very potent algae that have incredible detoxification and health properties. Spirulina contains the highest concentration of protein on the planet (roughly 60 percent complete protein), is rich in chlorophyll (which oxygenates and alkalizes the blood), and has immune supporting properties. It also contains DHA and high levels of vitamin A, which is great for your skin.

Among other amazing benefits, chlorella is one of the only foods that binds with toxic chemicals and heavy metals (like mercury) and moves them out of the body. It's one of the most cleansing things you can consume.

Fish Oil (or Algae Oil)

As I mentioned in the chapter on foods, adding a fish- or algae-based omega-3 supplement to your diet is one of the best things you can do for your health. I've added them to this section on blood support because among their benefits, omega-3 fatty acids are very important for cardiovascular health. Their anti-inflammatory properties reduce free-radical damage in the bloodstream and prevent LDL cholesterol from running rampant. All of that keeps your heart and blood happy.

Just as with greens powders, you'll find fish oils of drastically varying quality. This can be a lengthy discussion, so I've included one of my special health reports that goes into further detail on this topic as a bonus to this book.

In the meantime, here are some quick nuggets of information to help you choose a fish oil that's up to scratch:

Your oil should be:

- Derived mainly from small fish such as sardines, anchovies, and herring
- Molecularly distilled, thus ensuring removal of all toxins
- In triglyceride (TG) form, not ethyl ester (EE), as this ensures better absorption
- A source of at least 1,000 mg omega-3 per serving (with at least 400 mg of DHA)

If you'd prefer to choose an algae oil, then that's fine as well. Just remember that the above criteria won't necessarily apply to algae. Nonetheless, look for an algae oil that is as clean as possible and also contains a good amount of omega-3s (as listed above).

Once you've chosen your fish or algae oil, remember to keep it stored in a dark bottle in your refrigerator. This is of crucial importance because these omega-3 fats are extremely sensitive to light, heat, and oxygen, and should avoid contact with those three elements whenever possible. These same rules apply whether you're taking the oil in liquid or capsule form. Personally, I take fish oil in liquid form because it feels easier to digest and absorb.

Digestive Support

Betaine HCl

One of the common themes we've looked at so far is that low stomach acid is a big problem for almost everyone, which has resulted in a disordered-digestion epidemic. Thankfully, there's a supplement, betaine HCl, which naturally increases the level of hydrochloric acid in your stomach to help you better digest your food, especially protein. When choosing a good product, look for one that combines betaine HCl with pepsin (the enzyme that digests protein).

Before going any further, I need to warn you that although it's generally safe, not everyone should supplement with betaine HCl. If you decide to give it a try, you should always follow your doctor or natural practitioner's advice. Those at risk for complications are people who use anti-inflammatory drugs like corticosteroids, aspirin, ibuprofen (Motrin/Advil), or other nonsteroidal anti-inflammatory drugs (NSAIDs). That's because these drugs can damage the lining of your digestive tract and supplementing with betaine HCl could make things worse by increasing the risk of internal bleeding or ulcers.

Dosage? Take the Acid Test

Assuming you're good to go, you'll want to know how much betaine HCl to use. First of all, the best time to take HCl capsules is before your meal, which gives your stomach a few minutes to become more acidic in anticipation of the food to come. What you're looking for, odd as it may sound, is a slight burning sensation in your stomach, as that signals that it's now loaded with acid (which is what you want). However, please don't follow this protocol if you have any stomach ulcers, as excess acid could make them worse. Assuming you don't have any ulcers, here's what to do:

1. Start your protein-based meal by taking one pill (500–650 mg) of betaine HCl with a little bit of lukewarm water.
2. Stay at the one-pill dosage before your protein-based meals for another day, and if you still don't notice any burning on the third day, try two pills at your next meal.
3. Stay at two pills before protein meals for another day and then try three pills on the following day (day four) if no burning is present.
4. Keep increasing the number of betaine HCl pills taken with each protein meal until you notice some digestive discomfort/burning. When this happens, you will have reached your limit, and your ideal betaine HCl dosage would be one pill less. For example, if you felt the burning discomfort at four pills, then three pills would be your proper dosage.

Digestive Enzymes

I've already mentioned the importance of enzymes several times throughout this book. Scarily, because most of the foods we eat in their cooked and processed states are devoid of enzymes, we are ingesting fewer than ever before. Our dwindling energy and health is a reflection of that.

Aside from eating more fresh plant foods, the best way to get more enzymes is by supplementing with a high-quality digestive enzyme that will help your stomach and intestines to actually digest and absorb the food you eat.

For best results, take one to two digestive enzymes before, during, or after your meal. The best digestive enzymes will contain a wide spectrum that works on carbohydrates, fats, and protein. Examples of helpful enzymes to look for on the supplement label include: amylase, protease, lipase, lactase, phytase, cellulase, bromelain, glucoamylase, and hemicellulase. Trying to figure out the dosage within a digestive enzyme is nearly impossible, so don't worry about the units on the ingredient list; they seem to differ from one brand to the next.

There's also another perk to taking a digestive enzyme that has nothing to do with digestion: doing so in a fasted state (that is, between meals) has been shown to have numerous health benefits, such as reducing inflammation in the body, modulating the immune system, fending off viruses, and cleansing blood.

These effects can be beneficial for anyone dealing with allergies, food sensitivities, autoimmune problems, weight issues, and a host of other concerns. Enzymes are adaptogenic, which means they aim to restore balance in your body; it's ultimately what your body wants.[1] For this reason, try taking one to two pills or capsules throughout the day on an empty stomach, in addition to those you take with your meals. In most clinical settings, the doses are usually much higher, but I'm sure a few enzymes here and there can make a big difference for your health over time.

Probiotics

Most people associate the word *bacteria* with something bad; however, good bacteria are crucial to the existence of life itself. In fact, there are ten times more bacteria inside and around your body than you have human cells. Crazy, right?

What we're concerned with are the billions of friendly bacteria that live in your colon. They're known as *probiotics,* and you want them populating your large intestine to a greater degree, as they are crucial to the foundation of your immune system. Sadly, the overprescription of antibiotics these days is reducing the amount of probiotics each of us has and making us susceptible to more and more diseases.

Dysbiosis, the imbalance between healthy and unhealthy bacteria in your gut, is very common in today's society. It's not a surprise when you consider how much we abuse our digestive system through poor food choices, antibiotics, and chlorinated water. If you ever experience intestinal gas, bloating, stomach upset, allergies, or constipation—and who doesn't?—then you can definitely benefit from probiotics.

Aside from being an integral foundation for a healthy immune system, probiotics serve to complete the digestion of undigested foods, keep pathogenic bacteria and yeasts in check, add bulk to your stool, and produce vitamin K and some of the B vitamins.

Probiotics are naturally found in yogurt and fermented foods like sauerkraut, kefir, and kombucha—just to name a few. However, since most people don't really eat many fermented foods nowadays and because I don't recommend relying on dairy (yogurt), your best bet is to find a probiotic supplement that you can take on a daily basis.

Not surprisingly, there's a huge array of probiotic supplements on the market, many of which are a waste of your time and money. I'm going to make choosing a quality probiotic a little easier for you with the information below.

First of all, because there are many types of good bacteria—each with their own unique benefits—your goal should be to find a probiotic that contains a number of different bacterial strains and provides each in adequate quantities.

The following table gives you seven of the most important bacterial strains to look for in a good probiotic:

Bacteria	Main Benefits
Lactobacillus acidophilus	Digestive support, cholesterol regulation, antimicrobial
Lactobacillus rhamnosus	Diarrhea relief, anxiety reduction, weight management
Lactobacillus bulgaricus	Lactose intolerance prevention, diarrhea relief/prevention
Lactobacillus plantarum	Digestive and cardiovascular support
Lactobacillus casei	Leaky-gut prevention, cardiovascular health

Bifidobacterium longum	Digestive support, weight management, anti-carcinogenic
Bifidobacterium breve	Digestive support, enables other healthy bacteria to thrive

For best results, choose a probiotic supplement that contains anywhere from six billion to ten billion microorganisms (CFU) and take the dosage once or twice per day. Good probiotic supplements will also often contain adequate amounts of probiotics to promote the proliferation of healthy bacteria. It's best to take your probiotic on an empty stomach to allow it to pass through to the colon unimpeded.

Incorporating probiotics into your diet will make marked improvements in your digestion, elimination, and overall health. You'll notice your bowel movements improving in consistency and regularity, you'll help lessen the risk for the development of allergies, and you'll dramatically improve the health of your colon and the rest of your body. These friendly bacteria are truly amazing.

Adrenal and Stress Support

The following supplements can be very important for you if you suffer from adrenal fatigue or have chronically low levels of cortisol. Remember, cortisol is important to your body in the right amounts. It's only when there's too much (chronic stress) or too little (adrenal burnout from too much stress over time) that health problems begin to manifest.

The first three supplements I recommend below are known as *adaptogens,* which means that they help your body better deal with stress and reestablish your point of balance (homeostasis). For instance, if you have high cortisol levels, adaptogens will help to lower them. If low cortisol exists, they will help to elevate it. The best part is that these adaptogens really have no side effects when used in recommended doses. Furthermore, many studies have shown that the effect of adaptogens in combination with adequate vitamins and minerals (from food or supplements) works synergistically to restore healthy adrenal (and HPA axis) function.[2, 3] That's pretty neat.

Maca (*Lepidium meyenii*)

I've personally taken maca on and off for years with great success. It's a native root vegetable of the Andes Mountains in Peru that has long been used in all manner of folk remedies in that region. In our corner of the world, it's most often found as a powdered supplement at your health-food store. Its many benefits include balancing hormones, improving energy and mood, and increasing male virility and fertility.

Numerous studies have documented the health effects of maca.[4] A 2009 review of the literature revealed that maca has favorable effects on energy and mood; may decrease anxiety; improves sexual desire; and increases sperm production, sperm motility, and semen volume.[5] Since stress instructs the adrenals to shut down the production of sex hormones, it's no wonder that many people who have low energy (due to stress) also have a lowered libido. A little maca could go a long way between the sheets.

In addition to its hormone-balancing properties and energy benefits, maca has been shown to significantly decrease the levels of bad cholesterols—VLDL (very low-density lipoproteins) and LDL (low-density lipoproteins)—in the blood. It also improves glucose tolerance and provides numerous antioxidants (like glutathione) that are vital for cellular detoxification and health.[6]

Considering that most people with low cortisol (perhaps you, too) have impaired blood-sugar control and are at greater risk for cellular damage, enjoying a little bit of maca each day could do much toward bettering your health. It does have a unique, earthy flavor that might be off-putting for some, but it generally goes down easiest when mixed with water or added to a berry-based smoothie.

Rhodiola (*Rhodiola rosea*)

Rhodiola has a long tradition of use in Eastern Europe and Asia due its effects on supporting the nervous system, improving work performance, reducing fatigue, and decreasing depression. Most commonly available in capsule or tincture form, rhodiola extracts can produce a number of favorable changes to neurotransmitters, nervous-system activity, and cardiovascular health. Perhaps of interest to students and creative types, rhodiola has been shown to improve cognitive function (focus and concentration), decrease mental fatigue, and reduce burnout.[7]

Rhodiola can also improve exercise endurance, which is great for those who just don't seem to have enough fuel in the tank during their workouts. Numerous studies have shown that acute rhodiola intake improved endurance-exercise capacity in young healthy volunteers.[8, 9]

As always, follow dosage instructions on the bottle. Generally, anywhere from 100 to 300 mg of standardized extract per day in divided doses is recommended.

Ashwagandha (*Withania somnifera*)

Ashwagandha has often been referred to as "Indian ginseng" and has been a popular remedy in Ayurvedic medicine for many conditions for thousands of years. It is one of the best health tonics and restorative agents you can find, and had been used to treat general debility, exhaustion, stress-induced fatigue, and insomnia long before modern medicine existed.

Scientific studies have proven its rejuvenating effect, particularly for the muscles and bone marrow.[10] A 2012 study in the *Journal of Ayurveda and Integrative Medicine* showed a significant improvement in VO2max (a measure of aerobic endurance) and time to exhaustion in elite athletes supplementing with ashwagandha versus a control group during cardiovascular exercise.[11]

Unlike Panax (Korean) ginseng, ashwagandha has been shown to actually increase relaxation, which can be helpful for those who suffer from excessive stress.

A safe daily dosage is anywhere from 300 to 500 mg divided throughout the day.

Licorice Root (*Glycyrrhiza glabra*)

Licorice root extract was once used to make the candy of the same name, but has since been replaced by anise and corn syrup. I mention this because I'm definitely not recommending that you eat more licorice candy. Its history in the confectionary industry aside, licorice root extract definitely has healing properties, but only for certain people; because it binds to and decreases the half-life of cortisol (resulting in more effective cortisol), it should only be used by those experiencing low levels of cortisol.

Chronic high levels of licorice have been known to raise blood pressure, so please check with your doctor before considering using licorice root extract if you have normal to high blood pressure already. However, if your adrenals are already burned-out, chances are your blood pressure will be pretty low.

A safe dose for those with adrenal fatigue (low cortisol and low blood pressure) is between one and five grams divided in a morning and afternoon intake. Remember not to take licorice root at night since it will keep cortisol elevated, which will suppress melatonin—that important neurotransmitter required for a good night's sleep.

B Vitamins

I include the B vitamins here because, as a group, they serve to support your nervous system and red blood cells—both of which are critical for dealing with stress and producing energy. You can generally get B vitamins from foods like whole grains, potatoes, bananas, lentils, chili peppers, beans, nutritional yeast, brewer's yeast, and molasses. As you can see, however, most of these foods don't fall within the parameters of the All-Day Energy Diet recommendations. For that reason, a good B-complex vitamin supplement can be helpful for reducing anxiety, dealing with stress, and improving your energy.

A common misconception about popular, energy-boosting vitamin B_{12} is that it can only be found in animal products, thus putting vegans at risk of experiencing low-B_{12} status. Although the latter may be true, the reality is that B vitamins are derived from bacteria, and, ultimately, animals must obtain vitamin B_{12} directly or indirectly from bacteria. Similarly, the bacteria in our gut produce some of the B vitamins that we ultimately absorb back into our bloodstream.

So, in addition to animal meats, fermented foods in which bacteria and yeast play an integral role are also a viable source of vitamin B_{12}. As such, kombucha, kefir, and sauerkraut can all be great B_{12} sources for vegans and anyone looking to reduce how much meat they consume.

If you do supplement with a B-complex vitamin, you may notice that your urine glows bright yellow after taking your vitamin. This is due to the presence of B_2 (riboflavin), and it's nothing to worry about. Remember, your B vitamins are water-soluble, and I'd be more concerned if your urine didn't turn yellow, as that would suggest you're taking a poor-quality vitamin.

My Daily Supplement Ritual

Like I said, I'm no fan of relying on pills or supplements to maintain your health, but in my journey toward recovery from adrenal exhaustion, they have proved to be a fantastic support. Here, I'd like to give you my daily supplement ritual, which incorporates all of the supplements I've just discussed.

Aside from my morning glass of water, I'm constantly drinking water (with sea salt) all day long, in addition to taking the supplements in the schedule listed here. Remember, this is in addition to eating according to the All-Day Energy Diet principles that I've outlined in this book; without this diet, these supplements may very well be useless to you.

Feel free to copy my schedule if works for you, but please remember to consult with your doctor before starting with any of these recommendations.

Morning
(Between 6 and 8 A.M. or whenever the kids start going crazy)

- 2 cups water with lemon and a pinch of sea salt
- My Energy Greens in a glass of water with a pinch of sea salt
- Small glass of water with 1 teaspoon maca powder
- One adrenal formula pill, which includes rhodiola and ashwagandha
- Small glass of water with licorice root tincture (about 500 mg)
- 1 tablespoon liquid multivitamin (which includes appropriate levels of B vitamins)
- 1 tablespoon fish oil

Afternoon
(Between 12 and 4 P.M.)

- Small glass of water with licorice root tincture (about 500 mg)
- One adrenal formula pill, which includes rhodiola and ashwagandha
- My Energy Greens in a glass of water with a pinch of sea salt
- 1 to 2 digestive enzymes on an empty stomach

Evening
(Before bed)

- 1 to 2 probiotic capsules (about 10 billion CFU) on an empty stomach
- 1 tablespoon fish oil
- Magnesium tea to relax my body

As you can see, I drink quite a bit of water just as a vehicle for getting some of these nutrients into my body. When you add in my daily juice or smoothie, the odd glass of water here and there, and eating plenty of fresh vegetables and fruit, it becomes pretty easy to stay hydrated.

I should also note that whenever I eat a heavier meal that includes protein, I will start with several betaine HCl capsules and one to two digestive enzymes. In looking at this schedule, it may seem like a lot, but it's really not that time-consuming. After the first few days, this becomes habitual and runs on autopilot.

■ ■ ■ ■

Chapter Eight

DIGESTION DAMAGE CONTROL (AND HOW TO AVOID THE DREADED FOOD COMA)

By this point in the book, it's my hope that your kitchen is beginning to look a little different than it used to. Instead of clogging up your cupboards, refrigerator, and digestive system with caffeinated, carbonated energy drinks and sugary treats of all sorts, I hope you're stocking up on wholesome, healthy foods that actually nourish you. It's the crucial first step toward unlocking the kind of sustainable energy that you're truly capable of.

That said, this wonderful new diet will be of little use to you if your digestion isn't in order.

As we've seen in the previous chapters, years of stress and bad eating can leave your stomach broken. If you're suffering from low stomach acid or gut flora, your body lacks the ability to break down anything you eat properly, no matter how healthy it is. Without fixing your digestion, all of the rich minerals and vitamins in your healthy meals will go to waste.

Now don't get too frustrated, and certainly don't go tossing your fresh broccoli and kale in the garbage. Even if you suspect that your stomach isn't in the best shape, you should know that getting your digestion going again isn't as difficult as it may seem. In fact, this process begins with one simple principle: *how* you eat is just as important as *what* you eat.

Let's face it; most of us eat on autopilot, squeezing in meals during our too-short lunch breaks, or even in our car after we've made a frenzied detour through a fast-food restaurant's drive-through lane. By the time we get home in the evening after a long day, dinner becomes the most robotic of rituals. Without our even realizing it, the very act of eating becomes tainted by the rush and stress that hangs over the rest of the day, and that's where our digestive trouble begins.

Think about it this way: how do you perform under high stress? Probably not as efficiently as you would otherwise. The same goes for your digestion.

If you recall, I pointed out earlier in the book that digestion is not a process that takes place entirely in your stomach. In fact, it begins the minute you put a bite of food into your mouth. As such, if you're eating on the move or in a hurry, you're compromising the process your body needs to start breaking down your food for its long journey through your stomach, small intestine, and colon. Doing this can lead to indigestion and, in the long run, to myriad health conditions that emerge from improper digestion and a resulting weakened immune system.

I realize that this may be a little hard for you to wrap your head around. After all, how could eating takeout Chinese on the couch while you watch your favorite sitcom seriously impair your digestion? Over the course of the next chapter, we'll take a close look at how your current eating practices are hurting you, and more important, you'll learn how to get rid of these bad eating habits so that sitting down for a meal becomes the nourishing, regenerative act it's supposed to be.

Step One: Appreciate What's in Front of You

The next time you sit down to a meal, there's something I'd like you to do: absolutely nothing.

Well, technically you *will* be doing something, but what you *won't* be doing is shoveling food into your gullet the minute it's placed in front of you. When your meal is ready, I'd like you to stop; look at it with appreciation; and take three long, slow breaths before you dig in.

It's okay if that sounds a little ridiculous to you. Then again, if you're a spiritual person, it might not sound ridiculous at all, even if you don't do it regularly. The realm of religion and spirit aside, what I'm basically asking you to do is to take a moment to appreciate the food that you're about to eat. By doing so, you're not only detaching yourself from all your worldly worries, you're also shifting the focus of your entire body and mind to the act of eating, in essence priming yourself for full and proper digestion.

It doesn't matter if you're not a spiritual person and don't believe in saying grace. Just this simple act can calm your body into a parasympathetic state that "de-stresses" your digestive process. It doesn't need to be any longer than five seconds of contemplation and appreciation.

Step Two: Be Mindful about Where You Eat

When was the last time you sat down and ate a meal at your dining table? That may sound like a silly question, but with the pace of living in our modern world accelerating by the second, eating has become something we tend to do while we're doing something else, be it driving the kids to school or working at our desk. It's almost always an afterthought, and when it isn't, it's an indulgence. Rarely is it the simple act of nourishing ourselves.

Consider this a continuation of step one in that you're trying to become more mindful of what you're eating. There are two reasons for this:

- **First**, eating while you're occupied doing something else means there's some small level of stress and hyperactivity at work inside you. Your attention is being directed elsewhere. This is a problem because this stress level—however minute—will slow down your stomach's production of hydrochloric acid. This limits how effectively you can digest your food.
- **Second**, when you focus entirely on your meal, you're able to acknowledge when you're full, feel satisfied, and don't need to eat anymore. When you're distracted, you're far more likely to overeat. This has definitely happened to me in the past. I can't tell you how many times I sat down on the couch to watch a movie with a bag of potato chips in my hands. Halfway through the movie, and much to my horror, I would have finished the entire bag without even realizing it!

By getting up from your desk, computer, or couch and setting aside time to do nothing but eat, you allow your body to properly digest your meal, and you're also likely to eat less than you might otherwise. Not only does that help keep you trim and in shape; it also prevents you from overburdening your digestive system with far too much food. Eating becomes far less taxing on your body.

Step Three: Eat Slowly

This step can't be overstated. It's a natural progression from the first two steps in that now you're approaching your food with a Zen-like calm, fully present as you take in your nourishment.

It's really as simple as it sounds: pick up your fork or spoon, take a morsel of food, and put it into your mouth. Taste it on your tongue and begin chewing slowly and thoroughly until you have a mushy ball of food inside your mouth. Swallow it, and repeat.

If you find it a bit ridiculous that I'm walking you through this, consider how most of us practically vacuum down our meals. It's almost as if we're in a race to finish! When we do this, we can quickly overburden our digestive system as we're rushing our food down faster than the body can process it. The same goes for drinking as well: take your time.

Besides, eating slowly is the only way you can truly enjoy your meal. If you have some delicious food in front of you, why would you want to rush through it? Give yourself a chance to savor your meal. It's a far more enjoyable experience, and a healthier one as well.

Step Four: Chew Your Food!

Do I sound like a broken record yet? I apologize if it seems like I'm stating the obvious, but it's important that you really understand that all of these things we take for granted—appreciating our food, chewing it properly, and so forth—are crucial to effective digestion. If there's one thing you take away from this little "tutorial" on eating, it's that you should take your time eating your food, every step of the way. This guiding principle is especially important when you're chewing.

The simple truth is this: if you want to digest your food properly, then you need to chew your food thoroughly. You may have heard this before. You can even find articles claiming it's necessary to chew each piece of food 27 times before you swallow. That's hard to say, because a piece of steak will require much more chewing than a bite of banana, but the basic point remains the same: chew, chew, chew!

Chewing is perhaps the most underappreciated part of digestion. When you chew a piece of food, you expose it to the enzymes in your mouth, which really kicks off the process of carbohydrate digestion. If you're eating something that is more protein based (like steak), where the carbohydrate enzymes won't have much of an impact, chewing properly is still critical. By chewing your high-protein foods into a fine mush, you allow more of the surface area to be exposed to the protein-digesting hydrochloric acid in your stomach once it finds its way down there.

In a nutshell, chewing your food completely makes it easier for your body to process it, so take your time and turn your mouth into a machine. Just think of your mouth as your body's blender. The more you *blend,* the easier it will be for the rest of your digestive tract to extract as many nutrients as possible from your food.

Step Five: Limit Drinking While You're Eating

Everybody seems to have an opinion on whether or not you should drink before, during, or after a meal. Let's set the record straight: drinking during your meals is a big no-no.

First of all, your stomach is naturally a very acidic environment, and it needs to remain this way to break down the food you eat. When you drink during your meal, you're diluting the acid in your stomach, making your meal tougher to digest.

Second, if you drink during or even after your meal, the food you're eating will impede the liquid from passing through unobstructed. When this happens, whatever you're drinking will just end up sitting in your stomach on top of the food you've eaten while it digests. That can lead to a bloated feeling that's exactly what you're trying to avoid. You want to feel energetic after you eat, not sluggish!

It's okay to take a few sips of water during your meal, but you certainly shouldn't be chugging water while you eat.

It is, however, quite fine to drink a glass of water before your meal. In fact, I recommend it. When you do this, the liquid stimulates specific hormones that tell your brain that you're not as hungry. This means you'll end up eating less. The result is more energy, and less weight gain. You can even take this tip one step further by adding some lemon juice or apple cider vinegar to your water before you eat. It's a really cool trick that I do with almost every meal.

Lemon has a tremendous number of beneficial properties that help in the digestive process. By simply squeezing the juice of half a lemon into your water and drinking it before your meal, you'll be boosting your liver function, which will help your body eliminate toxins more readily. Lemon water also stimulates your stomach to secrete more acid, which will supercharge your digestion.

Apple cider vinegar operates much the same way, with the additional benefit of improving your insulin sensitivity. This means the vinegar will reduce the amount of sugar your blood absorbs when you do eat. Don't use this as an excuse to go bingeing on cakes, pies, and doughnuts, but if you know you'll be indulging in a sugary meal, you can use this neat trick to reduce some of the "damage" it will incur.

Even if you don't add lemon or apple cider vinegar to your pre-meal glass of water, always make sure that it's lukewarm. When you go out to eat at a restaurant, it's common practice for the waiter to serve you a glass of ice water before you even order your meal. That's something to be avoided, as the cold water shuts down the gastric glands in your stomach, limiting the amount of hydrochloric acid it secretes to digest your meal. Next time, take your water without ice, and be sure to do the same when you're eating at home.

There is one drink that you can have during your meals that won't hinder your digestion, and might even improve it. That would be red wine, as it's full of enzymes that aid in digestion. Beyond that, red wine has been found to help prevent heart disease. Again, I'm not encouraging you to go on a bender by any means, but an occasional glass with dinner—one, and no more—can actually be quite helpful.

Step Six: Take Digestive Enzymes or Acid Supplements with Your Meal

For the record, I am not a supplement pusher. I generally believe that your nourishment should be found in a diet full of wholesome foods. That said, when you're attempting to recover from conditions as insidious as adrenal fatigue and weak digestion, there are definitely particular supplements you can take to help your healing along. We'll look at more supplements later in the book, but right now, you should know about the incredible benefits of digestive enzymes and acid supplements.

Once your digestive system starts to suffer, your body struggles to produce adequate amounts of enzymes that are needed to break down your meals. Because of this, you might want to consider boosting your enzyme levels up again by taking digestive enzymes in pill form with your meals.

Go to any health-food store and you'll find dozens of different brands on the shelf. What you want to look for is a full-spectrum digestive enzyme that contains lipases, amylases, and proteases, as these three categories of enzymes will get to work on nearly everything you eat—fats, carbohydrates, and proteins.

You'll find directions on the bottle that probably recommend taking two or three pills before your meal. My suggestion is to go about it a little differently. Take one or two pills before your meal to prime your digestion, and maybe another during your meal if it's particularly complex. Once you've finished eating, take one or two more to finish it all off. By doing so, you'll cover each stage of your digestion, ensuring that there are enzymes to tackle your food every step of the way.

Similarly, acid supplements can work wonders as well. Earlier in the book, I introduced you to hypochloridia, a state of low stomach acid that can sink your health very quickly. You can combat this by taking a capsule of hydrochloric acid before your meal, which will provide the necessary levels of acid to aid in your digestion.

Again, I'd like you to feel this out. If you purchase an acid supplement, observe the directions on the bottle, but feel free to take a few more capsules if you feel the dosage isn't doing the trick. You'll know the capsules are working when you start to feel a slight burning sensation in your stomach. Don't worry—as I mentioned, this is quite normal and indicates that the acid is getting to work as it should. You may need one capsule, or you may need five. It all depends on your digestive strength. With time, you won't have to take as many to get the job done.

Remember that improperly digested food can lead to a whole host of nasty conditions, like leaky gut and a weak immune system. I don't like taking pills myself, but if they can help get your digestion back on track, you'll be thankful for them in the long run.

Step Seven: Eat Less

When all is said and done, one of the best things you can do to improve your digestion is to eat less food.

I'm a firm believer in this. With the advent of cheap, processed foods, the human appetite has swollen to monstrous proportions. Most of us today simply do not need to eat as much food as we do, and don't even realize how weighed down and slow we have become. This isn't normal, and certainly isn't healthy.

Mind you, this isn't just my opinion. Scientific studies have proven this as well.

In his *New York Times* best-selling book *The Blue Zones,* author Dan Buettner examined different places around the world—such as Sardinia, Italy, and the islands of Okinawa, Japan—where there are significant numbers of people living past the age of 100. He wanted to find out what accounted for their longevity. There were many factors he discovered, one of which is that many of these people tended to eat smaller amounts of food.

Without question, the less you eat, the more energetic you will feel. Hands down. Remember, digestion draws heavily on your body's store of energy, so the more food you eat, the more of your energy will be directed toward digestion.

Now, I'm not saying you should starve yourself, as you need food as an energy source to begin with. What I'm suggesting is that you should start eating less with each meal. Take note of how full you feel as you're eating. As a general rule, you want to put your fork down when you're 80 percent full, as five minutes after that you'll probably feel truly satisfied. Stuff yourself and you might feel as if you need to be rolled away from the table. That's the feeling we want to dodge.

As diet and exercise science become more mainstream, experimental concepts and theories that were once limited to the fitness community are becoming popular with the average person looking to get in shape. One of them is the idea that eating five or six smaller meals a day is the best way to keep your metabolism churning effectively, thus leading to accelerated weight loss and energy levels. Don't believe it for a second.

The reality is that eating five to six meals a day is not natural. Your Paleolithic ancestors certainly didn't have this luxury 100,000 years ago, so why would it be effective for you?

The truth is that you should eat as you need to, no more and no less. Sometimes, that might mean not eating at all. Don't listen to those who would have you believe that occasionally missing a meal is bad for you. It is not. If you don't feel like eating one night, it's perfectly fine to have a cup of tea or a small snack. As you observe the principles in this chapter, you'll become more in tune with your body's needs, and you'll know just when you need to eat. Your digestive system does not observe a strict schedule.

I like to observe the principle of calorie cycling, which calls for occasional periods of fasting, or simply eating less food. By choosing not to eat for an entire day every now and then, you allow your body to relax and spend more of its energy on functions other than digestion, such as breaking down immune complexes or repairing damaged cells. Basically, you're giving it the space and time to do a little upkeep. This process is known as *autophagy,* and when it occurs, your body scavenges around inside itself to eliminate dead matter and rebuild itself as needed. It's harder for this to happen when it's overworked from digesting food.

Some people scoff at fasting, but several studies have shown that it actually increases your body's ability to burn fat. That's because if you don't have carbohydrates coming into your body, your fat will temporarily become your body's main source of fuel.

The basic idea here is that we want to take a more measured approach to how much food we eat. So often, we wolf down our food in an act of escapism, wanting to alleviate the stress and worries we face day to day. That often leads to emotional overeating, or gorging on food we know isn't good for us. We've all been guilty of this at times, so it's nothing to feel bad about.

However, you can take a more conscious approach to your meals from here forward. There's a strong possibility you'll realize you've been eating quite a bit more than you need to.

Optional: Food Combining

Although food combining is not a principle that I'm thoroughly committed to myself, it has done wonders for my clients and may for you as well. Consider it an option worth experimenting with to really optimize your digestion.

Proteins are primarily digested in the acidic environment of your stomach, while the alkalinity of your small intestine tackles the digestion of fats and carbohydrates. The principle of food combining holds that any meal that combines all of these food types will be a burden on your digestive system, as you're compromising the ability of each of your digestive environments to function efficiently.

Consider the ever-so-manly meal of steak and potatoes. Steak is a dense, heavy protein that is digested in your stomach. Potatoes, on the other hand, are carbohydrates that are digested in both your mouth and your small intestine. According to the premise of food combining, eating this meal would be a no-no. Neither your steak nor your potatoes would be fully broken down, because you're diluting the environment each food needs in order to be adequately digested. Is it any wonder, then, that a hearty meal of steak and potatoes often weighs you down afterward?

What's the solution? Basically, it involves being mindful of the foods you eat together. For me, that boils down to two basic principles: (1) don't eat heavy proteins and starchy carbohydrates together, and (2) always eat fruit by itself.

For example, when I have steak, I tend to eat it with a steamed vegetable, such as broccoli or Swiss chard, and maybe a small salad on the side. Similarly, I tend to eat potatoes, pasta, and other carbs with vegetables. Rarely do I eat heavy proteins and starches together. That way, they don't compete for digestive juices.

You can also approach this idea by sequencing the type of foods you eat. Let's think back to Thanksgiving dinner again. According to the principles of food combining, this meal would detonate your digestion, as it so often does. One way around this is to eat foods in sequence. For example, you could eat your roll first, followed by the cranberry sauce, then the mashed potatoes, and then your turkey. This would prevent all your carbs and proteins from being lumped together in your stomach. I have to admit that's a bit of a boring way to eat, but again, this principle is to be applied at your discretion.

The idea behind eating fruit alone is this: fruit is digested very quickly in the stomach, and when you combine it with other foods, which tend to take a bit longer to break down in your system, the fruit ends up sitting in your stomach and fermenting. The result is a bloated feeling that often leads to gas and burping. I'm quite familiar with this, as my father's family is from Morocco, where the Mediterranean tradition is to eat fruit as a dessert.

You can be a bit more lax with this principle when you're making smoothies. Blending acts as a form of predigestion, making it easier for your body to break down fruits and vegetables even if they've been combined. As such, food combining is less of a concern here.

As I said, this is not something I'm very rigorous about, but I do believe there's something to it. It's worth giving it a shot.

■ ■ ■ ■

Chapter Nine

ENERGIZING EXERCISE THAT BURNS FAT AND DOESN'T CRUSH YOUR BODY

Do you like exercise? Is it something you can honestly say you enjoy? It's okay to admit that you don't. For most people, the very thought of it is exhausting. It's boring, sweaty punishment that takes every last thing out of you. Is it any wonder that so many people have such a difficult time sticking to a workout plan and reaching their fitness goals?

It's not supposed to be so difficult. In fact, when it's done correctly, exercise is supposed to fill you with a tremendous surge of energy that improves your approach to everything in your life: your work, your relationships, and most important, how you think and feel about yourself.

That's right; exercise is supposed to make you feel good.

The problem is that the popular approach to exercise is completely wrong. Worse than that, it's bad for you. The reason it feels like you're killing yourself every time you step on the treadmill is because you're likely doing your body more harm than good.

There are more and more studies being done on the lengthy, strenuous ways we tend to exercise, and the findings overwhelmingly point to one thing: the way we typically work out isn't working for us.

The answer is to take a more effective approach to exercise that doesn't wear your body out. It may sound like a bother, but refashioning your workouts is a pretty easy process that will not only reduce the amount of time you spend in the gym, it will actually make exercising more enjoyable. Best of all, you'll get better results in a much shorter time, whether you're struggling to lose a little weight or trying to build some muscle.

This refined approach is also key to unlocking the all-day energy that will help you power through your days with new vigor. Exercise correctly and you simply won't experience that midafternoon slump you face every day.

Don't worry if you've identified adrenal fatigue as a problem you're facing. With some slight modifications, *you* can adopt this radical new protocol as well. I'll outline those changes at the end of this chapter. For now, let's reframe how you think about exercise.

Rethinking Cardio

I mentioned treadmills before for a very specific reason: walk into any gym and the first thing you're likely to see is someone having a miserable experience on a treadmill. He or she is probably not alone. Look around some more, and chances are you'll see these breathless, defeated-looking people on the elliptical machines, stationary bikes, and rowing machines as well. There isn't much intensity to what they're doing, and they don't have any focus. They're just putting their time in. Simply put, they're suffering.

This is what most of us consider exercise, and it's very, very ineffective.

A study performed by the University of New South Wales, Australia, took two groups of people and closely monitored how they exercised and what results they achieved. Group One was instructed to exercise in the manner I previously described: 40 minutes of cardio at just above medium-level intensity, three times a week for 15 weeks. Group Two took an entirely different approach. What they performed is called interval training. This required members of Group Two to exercise at full intensity for eight seconds, then rest for 12 seconds. They repeated these two steps for 5 minutes, and over the course of the 15 weeks, worked their way up to a full 20 minutes of intervals.

The results speak volumes. On average, Group One actually gained 1.1 pounds of fat over 15 weeks, while the members of Group Two lost an average of 5.5 pounds of fat.

This study isn't the only one to yield such surprising results. Another study done at the University of Colorado analyzed the difference between 15 weeks of interval training and 20 weeks of traditional endurance training. Once again, interval training proved clearly superior, but with an interesting twist: those who took on the endurance training burned more calories, but less fat. On the other hand, those who performed interval training lost a whopping average of 450 percent more fat than the other group!

Why is interval training so effective? It all has to do with the fact that you're spending more time exercising at maximum intensity. When you do this, you spend a great deal more time at your maximum heart rate than you would during a typical workout. It may sound a little intense, but that's what really gets your body to burn that fat you want to be rid of. Furthermore, these bursts of intensity will also stimulate the release of adrenaline from your adrenal glands. This in turn will flood your body with a euphoric burst of energy that is often referred to as a "runner's high."

It's easy to adjust your current workout to this protocol. Let's say you regularly go jogging in your neighborhood. Instead of maintaining the same speed during your jog as you usually do, you're going to start at a light pace for a minute or so before sprinting as fast as you can for eight seconds. You then slow down to a gentle jog or even start walking for another 12 seconds before you start sprinting again. You repeat this over and over again until you've hit 20 minutes, or whatever time you would have usually spent jogging.

It might take a little getting used to, but you'll soon find that you now look forward to your regular jog simply because it makes you feel so good! It will also help you lose weight much faster than before.

There's no set pattern that you have to stick to. You can vary the length of your intervals as you like. You may decide to walk for a minute and sprint for 30 seconds; similarly, you may choose to sprint for 10 seconds and jog for 40. As long as you follow your bursts of intensity with slightly longer recovery periods, you'll be successfully performing intervals.

How often do you have to work out like this? If you don't have any adrenal issues and you're in relatively good shape, you can aim for three times a week, for about one hour combined. Maybe you'll choose to do three days of 20-minute intervals. It's also fine if you do 10 minutes one day, 20 minutes on another, and 30 minutes on your third and final day. It's all up to you. You do want to make sure you don't spend hours each week doing intervals, however, as this form of exercise can be quite taxing on your body physiologically as well as psychologically. You need to give your body some time to recover.

Best of all, this training lends itself to any type of cardio exercise you may be used to. Feel free to try intervals on the elliptical machine at your gym, with the jump rope in your garage, or even out on your bicycle. There are no limits to when and where you can make this training protocol work for you.

It also doesn't mean that you have to abandon long-endurance cardio completely, especially if you enjoy it. I like to recommend using these longer, old-fashioned workouts as active recovery on those days when you're taking a break from your interval training. A longer, slower run or bicycle ride can be a great way to keep your body from getting stiff between your major workout days. It will also allow your body to flush lactic acid out of your muscles, helping you feel good throughout the week.

Why Building Muscle Helps You Burn Fat Faster

Let's go back to that imaginary gym I mentioned before. We've already spotted quite a few people struggling on the cardio machines, but what's happening further back in the weights section? Usually, it's filled with lean, muscled athletic types. Why are they there and not on the treadmills and exercise bikes? It's because they know a secret that most people don't: building muscle helps you burn fat faster.

As we explored before, the common thinking is that you have to do hours and hours of cardio to slim down, but the best way to minimize your body fat is to build muscle mass. This doesn't mean you have to become a hulking bodybuilder, as building even a little lean muscle will significantly help you get trim and in shape.

Why is this? It all has to do with your metabolic rate. It's the speed at which your body burns calories when you're just going about your day normally. In fact, your metabolic rate is responsible for 70 to 75 percent of the calories your body burns on a daily basis, so the higher it is, the slimmer and fitter you're likely to be.

This rate naturally decreases as you get older, and the only way to keep it elevated is to create and maintain lean muscle. It's essential to staying in shape and maintaining the all-day energy you're seeking.

Now this doesn't mean you have to become a muscle-bound weight lifter. It's best if you begin with some simple body weight exercises that you're probably already familiar with: push-ups, squats, lunges, and pull-ups (if you can manage them).

A great body weight-training routine is quite different from what's traditionally practiced with weights in the gym. Whereas the old-school approach might call for a dedicated day to focus on your chest and back, and another to focus on your core and legs, a solid body weight workout will focus on bigger movements that incorporate a lot of different muscles. The result is a muscle-strengthening workout that also provides a tremendous amount of cardiovascular benefit, so you're also burning calories and fat at the same time. Pretty neat, right?

The routine you're going to get started with is actually pretty similar to the interval training I've already shared with you. It's called circuit training. As with intervals, circuit training requires you to switch back and forth between periods of high-intensity exercise and extended recovery periods. The difference is you're focusing on movements that engage several of your muscles at once. For example, a simple circuit training using body weight movements might look something like this:

- 30 seconds of push-ups
- 30 seconds of rest
- 30 seconds of lunge walks
- 30 seconds of rest
- 30 seconds of planks
- 30 seconds of rest
- 30 seconds of side planks
- 30 seconds of rest
- 30 seconds of step-ups
- 1 minute of rest

Perform this entire routine two or three times, and you've completed a successful full-body workout. Doing this on a regular basis will have you looking and feeling better in a matter of weeks. The most beautiful thing about it is how very simple it is. You don't need a gym membership to do it. In fact, you can perform it right next to your bed when you wake up in the morning!

You'll quickly notice that it's also a demanding workout, one that gets you huffing and puffing and breaking a sweat a little ways in. That's a sign that you're unlocking some cardiovascular benefits as well. As time progresses and you become accustomed to this routine, try increasing the length of your exercise periods and shortening the length of your recovery time. For example, you can do each exercise for one minute and only take 15 seconds for your recovery periods. This modification alone will greatly intensify your workout and improve your results.

As with your interval training, you don't want to be excessive with this workout. Perform it no more than two or three times a week. Feel free to do it on the days when you're not doing your intervals, or double up on both workouts: you can follow 25 minutes of circuit training with 15 minutes of intervals for an awesome workout that takes you less than an hour.

Now some women may be a bit daunted by this routine, thanks to the widespread myth that women who do any kind of muscle-building workouts end up looking bulky and masculine. Nothing could be further from the truth. Women's bodies contain around 10 percent of the amount of testosterone that men's do, making it highly unlikely for a woman to get as pumped and shredded as a man would. It would take a significantly more intense workout and diet protocol for the average woman to end up looking like a female bodybuilder, so if that's your concern, you needn't be worried. This workout will help anyone—male or female—look lean and fit.

Stepping It Up with Weights

Once you've been doing the body weight routine for a few weeks or if you're already somewhat trained, you should definitely start incorporating weights into your circuit training. We call this *resistance training*. Don't let the dumbbells scare you, whether you're a man or a woman. This approach won't necessarily make you bulky. You'll simply be adding heavier weights to your circuit workout, ensuring that your strength, fitness, and all-day energy don't plateau or even crash.

Of course, weight lifting will help you build your muscles. What you might not realize is that it will also strengthen your bones. This is especially beneficial for women, as they face a greater risk of osteoporosis as they age. With stronger, healthier bones and muscles working together, your body will become optimized to tackle all your day-to-day tasks in a more efficient manner. Life will become easier.

It is important to make sure that you're using the right weights, however. You don't want to be using something too light, as you need to be lifting weights that you can't lift for more than about five to eight repetitions. Let's put it this way: you wouldn't be able to lift your ideal weight ten times. It would in fact be dangerous.

Again, don't let this scare you. Resistance training is a natural progression from your body weight training, and the format is very similar. As before, you'll be following periods of exercise with periods of rest, only this time, you'll be performing five to eight repetitions of each exercise with weights in hand. Instead of a push-up, you'll do a bench press. Instead of a squat, you'll do a weighted squat. Once again, you'll pause for 30 seconds of rest between each exercise, although this time you'll take two minutes of recovery between each set. The resulting workout will look something like this:

- 5 to 8 bench presses
- 30 seconds of rest
- 5 to 8 weighted lunge walks
- 30 seconds of rest
- 5 to 8 repetitions of weighted squats
- 30 seconds of rest
- 5 to 8 repetitions of weighted step-ups
- 2 minutes of rest

This is a more intense workout, so you definitely don't want to do it any more than two to three days a week. If you perform this routine just as much as you need to, you will feel more alive than you have in years. Overdo it, however, and you might feel even worse. Remember: two to three days, tops!

Once you've stepped it up to this level, I like to recommend yoga or Pilates as active recovery. Once you're lifting weights, your body might get stiff and sore on your off days. However, both of these activities will help you stay loose and limber. I'm a huge fan of yoga. I see no reason why you shouldn't do it every day if you feel like it. It's an excellent relaxation tool that reduces stress and rejuvenates your body and your mind. It's also a perfect way to keep your energy levels up and aid your recovery in between your workout days.

How to Exercise If You Have Adrenal Fatigue

You may be wondering how exercise is even possible if you're struggling with adrenal fatigue. If you're tired all of the time and intense exercise would only make it worse by further wearing on your adrenals, then how could pushing your body any harder possibly make things better?

The key is to exercise smarter, knowing that you're not currently capable of the more intense workouts that everybody else is tackling. Your exercise plan needs to give your adrenals time to rest, recharge, and ultimately reset. Even when your adrenals kick back into gear, you have to be careful not to exercise too intensely, as you might blow them out again. Actually, that's something everybody has to be mindful of, even those who've never suffered from adrenal fatigue before.

As I've outlined throughout this chapter, the way that most people exercise is not optimal and can often be harmful. It's simply too much. Exercise tends to be treated like a last-ditch effort to save you from being overweight, when it should be seen as something you do to feel and look great! What's worse, most people in the fitness world don't consider that some people are suffering from adrenal fatigue and simply aren't capable of the insanely intense workouts that are so popular these days. I know this because I experienced it firsthand.

I burned out early. When I was a professional soccer player in my early 20s, my life was one big blur. I pushed myself ridiculously hard, and I didn't have a choice. Between games on the road and constant training, I blew past my physical limits six days a week, with one measly rest day to recover. It was way too much, but it didn't really catch up to me until I was in my early 30s. By that point, I couldn't avoid the truth any longer: unless I changed something, I was going to completely break my body. I was literally killing myself, one workout at a time.

The workout I'm sharing here is a distillation of all I learned during my difficult, yet successful, recovery. It's an exercise plan that builds on the principles I've discussed in this chapter, with an allowance for more rest time every step of the way. The resulting approach actually resembles a workout plan that power lifters and strength athletes might use, with a focus on building a lean, muscular frame instead. Best of all, it reduces the insanity (I mean *intensity*) of your workouts, which can help bring your adrenals back to life, supercharging your body with a reliable source of energy with which to power through your days.

■ ■

Let's begin by looking at your strength training. This workout is somewhat similar to the advanced circuit workout I mapped out in the previous section, and requires you to use weights. Once again, it's built on a foundation of only a few exercises: bench presses, lunge walks, pull-ups, and squats. The difference here, though, is that you'll be giving yourself additional time to recover between each exercise. The result looks like this:

- 5 to 8 bench presses
- Rest
- 5 to 8 weighted lunge walks
- Rest
- Pull-ups (maximum reps or 5 to 8 weighted)
- Rest
- 5 to 8 repetitions of weighted squats
- Rest

(Repeat three times.)

You'll notice that I haven't listed a specific period of time for your rest, and that's because there isn't a set time you should be resting for. It's all determined by how long you need to recover. With adrenal fatigue, you need more time to recover from exertion than the average person, and the best way to determine exactly how long that will be is simple: you have to allow yourself to catch your breath.

You may need 30 seconds to catch your breath and get going again, or you may need a full minute. However long it takes for you to stop huffing and puffing after each exercise is exactly how long you need, and you shouldn't feel bad about it. Once you've been doing this routine for some time, you'll need less and less time for recovery.

There's also another approach you might like to try called *superset training.* It calls for you to take two exercises that rely on different and noncompeting movements and muscle groups and perform them back-to-back. For example, squats use the pushing muscles of your lower body, and bench presses use the pushing muscles of your upper body. As such, they're perfect complementary exercises to put together in a superset.

Let's stick with this example. To perform a superset using these two exercises, you'll perform five to eight repetitions of your bench press, followed by some rest. Next you'll do five to eight squats followed by another rest period. Perform this about three times and you've just completed a superset. You're now ready to move on to your next superset of exercises.

It its entirety, a superset exercise would look something like this:

Superset 1

- 5 to 8 bench presses
- Rest
- 5 to 8 weighted squats

- Rest
- 5 to 8 bench presses
- Rest
- 5 to 8 weighted squats
- Rest
- 5 to 8 bench presses
- Rest
- 5 to 8 weighted squats
- Rest

Superset 2

- 5 to 8 step-ups
- Rest
- 5 to 8 pull-ups
- Rest
- 5 to 8 step-ups
- Rest
- 5 to 8 pull-ups
- Rest
- 5 to 8 step-ups
- Rest
- 5 to 8 pull-ups
- Rest

Both of these approaches create an exercise routine that pushes your muscles to maximum intensity, but they also give you the recovery time you need to keep going. Whichever one you choose, you'll be exercising in a fashion that will whittle you into shape while sparing your adrenal glands from the "go-go-go" intense workouts.

■ ■

The other component to your workout plan is, of course, cardio. Again, this is very similar to the reformulated cardio intervals I outlined earlier in the chapter. It's almost exactly the same, actually; the only difference is that you'll be doing it a little less.

To recap, we're doing away with the focus on excessively long endurance cardio. It's not good for anyone, and it's especially bad if your adrenal glands are shot. The preferred approach is *interval training,* which calls for short bursts of exercise followed by gentler rest periods.

The approach is simple: whatever your chosen method of cardio exercise is, you'll be breaking it up into short periods of maximum exertion and rest. If you jog, you'll split your jogging session into bursts of sprinting and longer periods of walking or jogging. If you jump rope, you'll split your session into spurts of rapid jumping and longer periods of slower jumping.

If you take a look back at the beginning of the chapter, you'll notice that I'm asking you to perform almost exactly the same cardio workout as I outlined for people without adrenal burnout. The key difference here is that you can't perform it as often, and you want to do it for much shorter periods of time. Your workout shouldn't take more than 20 minutes, and you don't want to do it more than twice a week, tops. Anything more than that will be pushing you beyond what you're capable of right now.

Don't worry if this doesn't seem like enough exercise. It absolutely is. In fact, even ten minutes twice a week is more than enough. The most important thing is that you keep moving just enough to tone your body while providing much-needed breathing room for your adrenals to come back to life.

In fact, that's a principle that applies to everyone, whether you're suffering from adrenal fatigue or not. Exercise is not punishment for a bad diet, and the old idiom "no pain, no gain" isn't relevant either; you don't need to exhaust yourself.

■ ■

When you discard all of the common fitness myths that so many people buy into, what you're left with is this: exercise is simply moving your body as it wants and needs to be moved so you can be the best possible version of yourself. In fact, the only time your body won't be moving is when you've passed away, so why not use it and enjoy it to its fullest capacity now?

Once you're exercising properly, you'll find that it doesn't leave you tired and drained. Of course, it will be challenging, but soon you won't be able to live without your workouts. They'll become a crucial part of the all-day energy you've been so desperately in need of. Imagine that.

■ ■ ■ ■

Chapter Ten

9 WAYS TO EASE STRESS AND BALANCE YOUR HORMONES

When your boss is putting the pressure on you at work, the last thing you feel is cheery and full of life. If you've been having constant arguments at home with your loved ones, you gradually become more and more miserable. If money has been tight and you've been struggling to pay your bills, it probably feels like the end of the world.

Stress is real and seemingly never ending in modern society. If it isn't one thing, it's another. As you've seen throughout this book, chronic stress is one of the most common reasons why so many people suffer from low energy. No matter what's stressing you out, chronic stress does quite a number on your adrenal glands and blood-sugar levels, and that's just the tip of the iceberg.

Remember, nothing in your body happens independently of anything else. All of your bodily functions are connected, making one ecosystem. It's impossible to only cleanse your liver without cleansing most of the other cells and organs in your body. It's also unrealistic to simply focus on fixing just one broken part inside your body. It's much the same with your life.

There will always be something that will confound, frustrate, and irritate you, and unless you can find an effective way to deal with the stress that arises, your health and energy will suffer. For that reason, this chapter will give you nine amazingly powerful strategies for easing your stress and giving some much-needed love to your adrenals. By finding a reliable way to reduce and manage your stress, you'll also experience incredible boosts to your thyroid function, libido, blood-sugar levels, and most of your hormones in general. Committing yourself to easy, consistent stress relief is one of the most powerful ways to bring your body back into balance.

I should note that these strategies are not nutrition related since we've already covered food and supplement strategies. Your goal should be to incorporate the crazy-powerful strategies in this chapter with the All-Day Energy Diet nutrition principles and adrenal-supporting supplements you've learned about already. Do so and you'll feel like a million bucks in just a few days.

Stress Management Isn't Sexy

For years, you've heard that stress management is important, but unfortunately, the recommendations often aren't as sexy as you might like. Who has time to commit to a new practice? Isn't there just a pill that will fix everything?

The reality is that sexy supplements and quick fixes don't win the race. It's those things we've all been told are good for us—yoga, meditation, spending time in nature, and so on—that are the true magic bullets. Of course, they take a little getting used to, but once you devote yourself to even one of these techniques, you'll experience the ease of being that has eluded you for so long.

We'll look at some specific strategies in more detail in a moment, and I'll also bring some lesser-known ones to the table. Implement any combination of these and you'll do your adrenals and your energy levels a huge service.

The Truth about Stress

Why is it that thousands of people sitting in a sports stadium watching the exact same game experience it completely differently? Why is it that the sound of a crying baby on an airplane drives some people crazy, while it triggers empathy in others?

The answer to these questions can be summed up in one word that's responsible for all of the stress you will ever experience: *meaning.*

There is no situation or event that elicits any particular response in us; rather, it is the meaning we give to that situation or event that creates the emotions (that is, stress) we experience. If your favorite football team is down by a few points with just seconds to go in the game, you'll be a little more stressed-out than the fan whose team is winning. A baby crying on an airplane is not an inherently stress-inducing situation. How you respond to the baby's crying is completely up to you; it has nothing to do with the baby. If it were the baby's fault, then every single person would respond the same way, but we don't.

Knowing this is very important and empowering because it means that you are in full control of your emotions. If you're stressed-out, it's your fault. However, if you're happy and serene most of the time, that's your fault, too! When you understand that everything in your life is your responsibility, you then have the power to take control.

In fact, speaking of control, many studies have shown that we tend to be more stressed and less happy when we feel less in control of a specific situation. A classic example of this was presented in the *Journal of Personality and Social Psychology* in 1970.

In this study, 40 male undergraduates were told to react to the onset of a six-second shock. Following ten trials, half of the subjects were told that by decreasing their own reaction time they would reduce the duration of the shock. In essence, they were given *control.* The other half of the subjects were simply told that the shock duration would be reduced, thus no control was imparted to them.

All 40 subjects, regardless of group assignment or reaction time, then received three-second shocks in the second half of the study. During this phase, subjects who believed they had control (by reducing their reaction time) showed *lower* stress response to the shock than subjects who felt they had no control, indicating that perception of control can affect autonomic responding (that is, stress response).[1]

A more recent study in 2004 showed that when participants had more control over a noise stressor, they had a reduced cortisol response when the stressor was presented.[2]

Results like these have been demonstrated time and time again. What do these kinds of findings mean to you? I hope what you're getting is that accepting more responsibility for your life can be very empowering in terms of how you experience it. Only you can control your emotions.

Here's something else to think about: many people are scared to fly. I'm not one of them. I love flying, but I believe that a lot of people are afraid of flying because they don't know what's happening in the flight deck. If you sat behind the controls of an airplane and realized just how safe it was (unless you did something incredibly stupid), you would feel much more secure and less stressed the next time you sat as a passenger on a commercial airliner.

The reality is that flying is statistically safer than driving, yet few people freak out about driving. According to the National Highway Traffic Safety Administration, there were more than five million vehicle accidents in 2008, yielding 1.27 fatalities per 100 million vehicle miles traveled.

By contrast, the National Transportation Safety Board reported only 20 accidents for U.S. airliners in 2008, which represents almost zero accidents per million flying miles. No one died in these airplane accidents, and only five people were seriously injured.[3]

If these numbers aren't reassuring enough, then check out these odds of dying from the National Safety Council from 2008—the relative risk of dying in a motor-vehicle accident is 1 in 98 for a lifetime. For air travel, the odds are 1 in 7,178 for a lifetime. Wow!

Just looking at those odds, you'd almost be crazy to get into a car, but flying seems more dangerous because the risk perception is so much greater. Driving gives us more personal control, so it feels less dangerous. Plane crashes can also kill more people at once. Naturally, these rare events conjure up massive media attention because the media loves to scare us, making more and more people sensitive to, and even apprehensive about, flying.

I mentioned that I love flying. In fact, not too long ago, I pursued my private pilot's license. For me, flying is more relaxing than meditation. Seriously. When you're flying a small plane, you don't have time to think about anything else other than the task at hand. There's no time to check your e-mail or to worry about work or your bills. If you aren't present, you risk getting seriously hurt and potentially hurting others. Now, I'm not saying you should become a pilot; I'm just providing some perspective.

Let me illustrate this with one more example, this time from sports. I was fortunate to have played professional soccer for three years in my early 20s. However, my demeanor on the field was far from adrenal friendly. If you knew me in person, you would have thought I was possessed by a demon while playing soccer. As a goalie, I was a tyrant on the field—yelling and directing my teammates for 90 minutes. Although I wasn't running from end to end, I was exhausted after most matches, and drained near the tail end of my career. I was only 24! At the time, I had no idea, but I had crushed my adrenals by the amount of emotional stress I subjected my body to. Couple that with pregame jitters, intense training, and a poor diet, and it's easy to see why I lacked the energy I needed on a day-to-day basis.

Nowadays, when I play soccer, tennis, or anything else, I'm much calmer and find ways to smile and enjoy the experience. The difference is night and day. In the past, if I made a save, I would yell mercilessly at my defenders for allowing the shot to come through. Now, I make the save, get up, and have a nice chuckle—most of the time.

Again, the situation is still high intensity with lots at stake, but I've become more aware and more present to recognizing what's happening. You can't fight fire with fire. You need to become the calm in the eye of the storm. That's actually my mantra when life gets crazy. And I especially repeat this when my little kids are going nuts.

Based on the conditioning I received from my parents, my default setting is to react in a not-so-great way when my kids are yelling or crying. In coming to realize that tendency, I've become much more aware that *I am not my parents.* How I respond is my choice—not that of my conditioning.

So, when the kids are going nuts, I repeat to myself, *I'm the calm in the eye of the storm,* and visualize myself meditating in the eye of a crazy hurricane. It's amazing how powerful that simple exercise can be. Pair that with a few deep breaths and you're back in full control, no longer reacting to whatever "should" be stressing you out.

I use these examples to illustrate that stress is 100 percent self-imposed. Some people freak out when something happens, while others find a different and more empowering meaning. All else being equal, those who freak out will most likely wear down their adrenals and health more readily than the Buddhist monk.

Now that you have a better sense that your stress is on *you,* let's look at some of the best ways to manage your state more effectively so that you give your adrenals—and hormones—some much-needed support. Remember, you can't eliminate stress; you can only handle it more efficiently. The good (and bad) news is that it's completely up to you.

Mindfulness and Presence

How you choose to look at life can make all the difference in how you experience it. It's all a matter of perspective. Some people drift through life feeling like bad things always happen to them, while others take a little more responsibility by choosing to believe that they are the cause. The former is choosing to live life as a victim, where you essentially hand over your personal control to fate. Choosing to believe that you create what happens in your life is a more empowering position to take.

Whichever belief you live by, this remains true: it's how you define an event that determines how your body responds to it.

With that said, I encourage you to become more present and aware, and more conscious about things that occur in your life. Is sitting in traffic really the end of the world? If your flight is delayed because of mechanical problems, should you really be freaking out, considering the potential disaster that could occur if it were to otherwise take off?

I know what you're thinking: you can't just flip a switch and eliminate those bothersome thoughts. I know it's tough and I'm not saying that you'll experience a miraculous breakthrough overnight, but I want to challenge you to become a better, more conscious version of yourself. I want you to be someone who can experience an event and attach an empowering meaning to it that actually serves you, instead of one that stresses you out.

In any given situation, ask yourself these empowering (and stress-reducing) questions:

- *What can I learn from this situation?*
- *How can this situation or event serve me?*
- *What does this situation mean to me, and how can it help me move forward?*

Gratitude Journaling

Your brain can only focus on one thing at once. That includes emotional states. You cannot feel sad if you are smiling. Likewise, you cannot feel stressed when you focus on what you're grateful for. Therefore, the easiest and most powerful activity you can start doing right now to relieve stress and increase your happiness is to spend just five minutes writing down what you are grateful for. It doesn't matter if you do it in the morning before work, or at night before bed. You can even do both. The most important thing is that you commit to it.

Remember, what you focus on in your life expands. You can only move forward by focusing on what's working. Compare that to constantly complaining and focusing on what's *not* working. I bet that even by thinking about this, you're beginning to feel the difference already.

Here are a couple of questions to give you some guidance in your journaling:

- *What are three things I'm grateful for?*
- *Who are three people I'm grateful to have in my life, and why?*
- *What were three wins I had today?*

That's it. Just start with those. You don't have to make it more complicated than that. These three questions will instantly shift your focus and get you on the path of gratitude and greater confidence. Ask better questions and you'll automatically get better answers.

Deep Breathing

What's the one thing you can't live without but probably take for granted every second of the day?

Air (oxygen, specifically). Go without it for more than a few minutes and you're done for.

Breathing brings energy-producing oxygen into your body. Without oxygen, you lose energy. Without oxygen, your body becomes more toxic. Without oxygen, you eventually die. When you breathe more fully from your abdomen, rather than taking shallow breaths from your upper chest, you inhale more oxygen. The more oxygen you get, the less tense and less anxious you feel.

Here, I want to share two unique breathing techniques with you that will deliver more oxygen to your body, while helping to calm your nerves tremendously. With both breathing techniques, make sure that you sit with good posture and keep a long neck (chin not tucked in toward your chest) to allow optimal airflow into your lungs.

Four-by-Four Abdominal Breathing

The key to this breath pattern is to inhale deeply from the abdomen, getting as much air as possible into your lungs.

1. Sit comfortably with your back straight. Put one hand on your chest and the other on your stomach.
2. Breathe in through your nose for four seconds. The hand on your stomach should rise. The hand on your chest should move very little.
3. Hold that breath for four seconds.
4. Exhale through your mouth for four seconds, pushing out as much air as you can while contracting your abdominal muscles. The hand on your stomach should move in as you exhale, but your other hand should move very little.
5. Hold that exhaled position for four seconds.
6. Repeat this pattern ten times.

If you can commit to doing this quick two-minute breathing exercise anytime you feel stressed, you'll be much better equipped to handle anything. At the very least, do this exercise once in the morning and once at night.

Alternate-Nostril Breathing

This breath pattern is meant to bring calm and balance, and unite the right and left hemispheres of your brain.

1. Starting in a comfortable seated pose, hold your right thumb over your right nostril and inhale deeply through the left nostril.
2. At the peak of your inhalation, close off your left nostril with your ring finger, and then exhale through your right nostril.
3. Inhale through your right nostril, closing it off with your right thumb and then exhaling through your left nostril.
4. Repeat ten times.

This is a great breath pattern to use before doing any kind of creative or focused work.

Yoga

Several years ago, when I was the strength-and-conditioning coach for men's soccer at the University of Toronto, I instituted a 25-minute yoga routine (which I now call Yoga for Athletes) two to three times per week with our players. At first, everyone thought I was crazy, but after just one session, some of the guys came up to me saying that their legs felt so much lighter and that they felt more focused and "in the zone." These were elite college athletes—not your typical yoga demographic. Within the course of a few weeks of adding in this new routine several hours before games, after training, and while traveling, I was getting feedback like:

- "My muscle soreness after games has been completely eliminated, and my agility on the field has greatly improved because of my increased flexibility."
- "I'm finally injury-free and more focused as I prepare for important games."
- "These yoga sessions helped me focus and perform better than I could have expected."

- "I felt calmer and less nervous heading into big games."
- "The tension I once had in my lower back is now gone."

Imagine what these kinds of benefits could mean in *your* life. I don't think I need to explain the numerous stress-relieving, body-rejuvenating, and energizing benefits of yoga. There are many types of yoga, so just find one that suits you. For best results, I suggest doing yoga at least two to three times per week. If you want to do it daily, then that's even better, especially if you're suffering from adrenal problems.

Meditation vs. Punching Away Your Anger

Very often, stress leads to anger. Interestingly, there is an assumption that anger and frustration build up inside of us and that "blowing off some steam" by punching a bag or yelling will make us feel better. We call that *catharsis* and assume that it's a viable way of reducing stress and anger. Unfortunately, it's not.

A study in the *Journal of Personality and Social Psychology* showed why this is so. In this experiment, the researchers had each participant write an essay on a sensitive topic and then told them that it would be evaluated by a peer, when, in reality, it was to be evaluated by the researchers.[4] A high-anger condition was then created, as each subject was told his or her essay was poor and was "one of the worst" they had ever read. Think back to a time when your parents or teachers told you your work was terrible. Not a good feeling, right?

Soon after, some of the subjects were given the opportunity to punch a punching bag for two minutes. Other participants weren't required to do anything. Then, everyone played a video game against a fictional opponent in which they could punish their opponent with blasts of noise. The loudness and the length of the noise were used as measures of aggression.

If catharsis were really true, then the subjects who had a chance to take out their anger on a punching bag would have been less angry than those who had to sit and do nothing at all. Amazingly, the opposite was observed. Those subjects who punched the punching bag were actually more aggressive than the people who did nothing.

The researchers concluded that punching a punching bag reinforces the link between being angry and acting in an aggressive manner. I would also add that such aggressive behavior would likely increase testosterone levels, which would fuel the aggressiveness even further.

Based on these results (and similar findings elsewhere), sitting quietly and meditating is a much more effective way of calming yourself down than attempting to let off steam through another aggressive act.

Countless studies have shown the benefits of meditation on stress and anxiety reduction as well as on reducing the risk of heart disease. I won't bore you with them here. Instead, I'd like to give you some helpful ideas for making meditation a regular practice, if it's not already in your life.

I'm no expert in meditation and I don't endorse one method as better than another, but I can say from experience that anything you can do to quiet your mind and just "be" will be extremely helpful for you. In my life, whenever I meditate, things just seem to go better. Time slows down, I feel happier and more relaxed, and business picks up—just to name a few observations. I'm not saying the same will happen for you, but you *will* experience some great things.

Meditation is like working out for your mind. Initially, just sitting still with your eyes closed for two minutes will seem like an impossibility. You'll fidget, and your mind will wander. That will lessen with practice. As with exercise, the more consistently you train your mind to be still, the easier it will become to achieve that state on a regular basis. As many meditators have experienced, you'll likely feel calmer, more focused, and less distracted. Much like a new computer with nothing taking up its hard drive and RAM, when you have less clutter in your mind, you tend to feel more energized and perform better. You really do win on all accounts.

If you've never meditated before, here's what I recommend you do to get started:

- Choose a consistent time and spot for your meditation.
- Commit to practicing daily.
- Initially, commit to just five minutes. You can always build from there.

I encourage you to sit in silence or at the very least use an audio of binaural sounds that will help you enter into the ideal brain-wave state for meditation.

Remember, there's no wrong way to meditate. Some may argue that there are more effective forms—which may be true—but any meditation that provides peace and tranquility will benefit you.

Now that you're all set to give it a go, start with this simple meditative exercise:

1. Sit in a relaxed position. You can either sit in a straight-backed chair with your feet flat on the floor or on a thick, firm cushion on the floor. *Note:* If you'd rather lie down to start, then just be careful that you don't fall asleep!
2. Keep your back, neck, and head vertically aligned. Relax your shoulders. Find a comfortable place for your hands (usually on your knees).
3. Bring your attention to your breathing. Observe your breath as it flows in and out in a natural manner. Give full attention to the feeling of the breath as it comes in and goes out. Whenever you find that your attention has moved elsewhere, just note it, let go, and gently escort your attention back to your breathing. That's it. Simply do your best to maintain

awareness of your body and breath, and when your mind wanders, gently bring it back to your breath. As with any new skill, this will become easier with practice.

Nature Therapy (Ecotherapy)

Before my kids were born, I had an amazing morning ritual. I would wake up early, pack the dogs in the car, and drive over to the local forest about ten minutes away. Once there, I would let them loose, and we would run through the wooded trails for about 30 minutes. I don't think I ever felt better in my life. The aerobic exercise, which flooded my body with oxygen, in combination with the beautiful crisp air from the forest, was truly revitalizing. By the time I returned home around 8:30 A.M., I felt like I could take on the world. It was terrific. In fact, I need to find a way to get that back into my routine, even with three little kids.

I've always been drawn to nature, and I love walking and exercising outdoors. Drop me off in a forest or near a lake and I'm good for hours. I don't think I'm the only one who feels this way. Numerous studies have shown the beneficial impact of walking in nature on mental and overall well-being.

In one study, participants were randomly assigned to either walk in a natural environment, walk in an urban environment, or relax in a comfortable chair. At the end of each exercise, the researchers found that those who had taken the nature walk had significantly higher scores on overall happiness and positive affect and significantly lower scores on anger and aggression. The nature walkers also performed significantly better on a cognitive performance measure.[5] I guess my experience wasn't unique.

Another study out of the University of Essex compared the benefits of a 30-minute walk in a country park with a walk in an indoor shopping center among mental-health patients. After the country walk, 71 percent reported decreased levels of depression and said they felt less tense, while 90 percent reported increased self-esteem. By contrast, only 45 percent of participants experienced a decrease in depression after the shopping center walk, after which 22 percent said they actually felt more depressed. Some 50 percent also felt tenser, and 44 percent said their self-esteem had dropped after window-shopping at the center.[6]

It's not surprising that nature has a therapeutic effect when you consider that humans have been closely bonded with it for all our existence. Only recently have we traded in the great outdoors for "boxed" indoor living. It's almost primitive, but contact with green spaces is like going back home for our DNA. I think one of the main reasons why nature can heal and transform us is because of its calming and mind-quieting effect. It's like doing a moving meditation. Think about it—in nature, our minds aren't bombarded with billboard ads and thousands of other distractions. There is less information to process. Much like a computer, the fewer "programs" we have running, the better we will perform.

The message is simple—get outside every single day and spend some time in nature. Take your dogs for a walk, go for a bike ride, try snowshoeing, or do anything that connects you to nature. Remember that any activity that declutters your mind and improves your mood will generally give you more energy.

Aerobic Exercise

In the chapter on exercise, I discussed how to work out if you want more energy. Now I want to hit on the importance of aerobic exercise, especially for its role in easing stress. But let's be clear—what I'm referring to here is slightly more intense than a relaxing walk. The great thing is that you can combine aerobic exercise with nature therapy, kind of like I described in my own life just a moment ago.

Aerobic exercise is defined as any exercise that is predominantly fueled by the presence of oxygen. It isn't overly intense, and it doesn't usually create a lot of lactic acid (which causes the burning sensation in your muscles). It can take the form of jogging, cycling, swimming, skiing, or anything else that raises your heart and breathing rate. If you're a technical person, this type of exercise would take place at around 60 to 75 percent of your maximum heart rate.

Why is aerobic exercise important for reducing stress?

It has a unique capacity to exhilarate and relax, to provide stimulation and calm, to counter depression and dissipate stress. It's a common experience among endurance athletes and has been verified in clinical trials that have successfully used exercise to treat anxiety disorders and clinical depression. If athletes and depressed patients can derive psychological benefits from aerobic exercise, so can you.

Most of the mental benefits of aerobic exercise have a neurochemical basis. Put simply, it reduces levels of the body's stress hormones, such as adrenaline and cortisol. It also stimulates the production of endorphins, chemicals in the brain that are the body's natural painkillers and mood elevators. As I explained before, endorphins are responsible for the "runner's high" and for the feelings of relaxation and optimism that accompany many hard workouts.

And let's not forget about the behavioral aspects that contribute to the emotional benefits of exercise. As your waistline shrinks and your strength and stamina increase, your self-image naturally improves. You feel better about yourself. Happier. More confident. It's a delicious snowball effect.

One of the best ways to start your day is by doing some aerobic exercise. It doesn't have to be long, but a few minutes of moderate-intensity exercise can awaken your body and mind more effectively than any cup of coffee. You don't have to kill yourself. Just start with 10 to 15 minutes of your favorite exercise. And to ensure that you're in the right intensity zone, just use the Talk Test. All this means is that you should be exercising at an intensity at which you can hear your breathing, yet you're still able to carry on a conversation.

And due to the lower intensity of aerobic exercise, it doesn't crush your adrenals like so many intense workouts do. With that said, I wouldn't base your entire exercise regime on aerobic exercise (for reasons I described in the exercise chapter), but it's an important component.

Tapping

Some time ago, I was introduced to a unique stress-management technique called *tapping* or *Emotional Freedom Techniques* (EFT). I instantly fell in love with it and applied it whenever I felt stressed. It is such a simple and powerful tool that everyone needs to know how to use it. For that reason I've called upon my good friend Nick Ortner, author of the *New York Times* best-selling book *The Tapping Solution,* to show you why it works and how to do it. Take it away, Nick . . .

> When Yuri told me he was writing a book on how you can rescue yourself from the chronic lifelessness that so many of us deal with from day to day, I got really excited. If there's one thing I've learned from helping thousands of people around the world overcome their personal challenges—everything from relationship issues and trouble at work to struggling with weight loss and even chronic pain—it's that you can improve and even heal almost any problem you're facing by reducing your elevated stress level.
>
> Yuri knows this, and I expect that by reading this book, you've come to believe it, too.
>
> Sadly, this type of thinking is still somewhat taboo in the mainstream health-care industry, although new studies emerge every day that challenge this stubborn position. Interestingly, this approach to healing has been a huge part of the medical traditions of the East for centuries and still is today. What I want to share with you is an empowering stress-relief technique that draws upon this ancient knowledge as well as the insights of modern psychology. It's the easiest and most powerful way that I know of to dissolve the stress that's been depleting your energy every day.

It's called Emotional Freedom Techniques, or EFT for short. Millions of people around the world simply know it as *tapping.*

What makes EFT so incredible is that much like acupuncture and acupressure, it addresses your body's energy pathways, known as meridians. By using your fingers to tap on select points along these channels and talking through your problems as you might in your therapist's office, you're able to release the energy blockages that have been creating such trouble in your body and your life. You're essentially clearing away the stagnant energy, thoughts, and outlooks that have been derailing you every day. You'll be astonished by how fast it can ease your tension and help you along.

For many people, this explanation isn't convincing enough; it sounds too airy-fairy and alternative, maybe even magical. Thankfully, modern science has been confirming what EFT practitioners have known for so long. Specifically, a study done at Harvard Medical School found that applying even slight pressure to the meridian points used in tapping significantly calms your body's stress response. As Yuri so elegantly outlined in Chapter 1, the stress response in your brain and your adrenal glands and the resulting flood of cortisol in your body is so often to blame for everything from sluggishness to depression. This landmark study shows how EFT can tame that tide of cortisol that's been hijacking your life.

Another study done by Dr. Dawson Church found that tapping could also greatly help people with the highest stress levels of all—war veterans. The study was conducted on veterans suffering with post-traumatic stress disorder, commonly known as PTSD. They were split into two groups, with the first receiving EFT coaching and traditional talk therapy, while the second group received talk therapy alone. The results said it all: only 4 percent of those who went through talk therapy by itself experienced reduction of their symptoms, whereas 90 percent of those who were exposed to tapping reported that they were no longer experiencing symptoms of the disorder. If it helped them, imagine what it can do for you!

If you're wondering why tapping should excite you more than the dozens of other approaches to stress relief to be found these days, it's because of how very easy it is. It's truly a do-it-yourself personal care technique, one you can teach yourself in a matter of minutes. You can certainly work with a trained EFT therapist, but you can also simply do it on your own when you wake up each morning. The choice is yours.

As with anything, the only way to know if tapping works for you is to try it for yourself. I'd like to give you a quick video lesson myself, which you can find on my website at thetappingsolution.com/what-is-eft-tapping. Give it a shot and I'm certain you'll become a fan, just like Yuri!

Sleep

Arguably, sleep is the most important component of good health. Poor-quality sleep has been associated with almost every problem, including diabetes, obesity, heart disease, and obviously, low energy. The thing to remember about sleep is that it's when your body does most of its healing, repair, and growth. That's why babies require so much sleep.

You can find all kinds of products to improve your sleep—from biohacking devices to sleep aids and more—but I'm going to keep things really simple for you: the human body is encoded with a natural circadian rhythm, which is based on the light and dark cycle. This basically means that when it's light outside, your body is primed to stay awake, and when it's dark outside, your body is ready for bed. As I mentioned earlier, cortisol is highest first thing in the morning (when the sun rises) and lowest at night (after dark). By contrast, melatonin (the sleep hormone) islowest in the morning and highest at night. It's nearly impossible to have both elevated cortisol and melatonin at the same time.

So think about this—if you're stressed-out before going to bed, will you sleep very well? Obviously not. The stress-induced cortisol coursing through your blood inhibits melatonin, which means you're going to have trouble falling and staying asleep. Thankfully, you can apply any of the previous stress-easing strategies before bed to alleviate this problem.

Modern living poses another problem for good, restful sleep: light pollution. Specifically, it's the blue wavelength of light that suppresses melatonin. Essentially, the longer and more intense your exposure to blue light, the more the melatonin in your body will be suppressed.

Blue light is most pronounced in artificial light sources; TVs, computer screens, and household lights are the main culprits. Fortunately, there are ways of reducing blue-light exposure, especially before turning in for the night.

First, shut off all lights and electronics at least one hour before going to bed. Sure, this might not be that feasible, but at least shut off your TV and computer and do some reading instead. That means from a real book; your iPad or Kindle won't cut it.

Second, try wearing glasses that block out the blue light. Blue-blocker sunglasses are great for this. Often, I'll wear mine at night if I'm watching TV. You might get some funny looks, but it can definitely help you sleep better. Researchers at the University of Toronto (my alma mater) compared the melatonin levels of people exposed to bright indoor light who were wearing blue-light–blocking goggles to people exposed to regular dim light without wearing goggles. They found that melatonin levels were about the same in both groups, showing that blue light is a potent suppressor of melatonin.[7]

Aside from wearing blue-blocker sunglasses and turning down your lights at night, let me give you some specific strategies to get a more restful night's sleep:

— **Go to bed and wake up at the same time every day.** Yes, even on weekends. This will help reset your body's internal clock, and within a week or two you won't even need an alarm clock. I recommend going to be bed no later than 11 P.M. and waking up naturally between 6 and 7 A.M. Remember, your body likes routine—kind of like the sun rising and setting every day.

— **Brain dump before hitting the sack.** If you have trouble falling asleep because you've got too much on your mind (stress) and thus too much cortisol flowing around, take five to ten minutes before bed and write down everything that's rattling around in your head. This could be things you're worried about, to-dos for tomorrow, or anything else. Having it down on paper frees your subconscious mind from having to worry about it. A free mind equals better sleep and more energy the following day.

— **Take an Epsom bath and/or have some magnesium tea before bed.** Epsom salt baths relax your body and can be a great way to unwind after a hectic day. Likewise, drinking magnesium tea one to two hours before bed can help calm your body since magnesium is the *calming mineral.*

Whatever you decide to do, please remember that the *quality* of your sleep is more important than the quantity of your sleep. I used to sleep 12 hours a night but never felt rested. Now, by incorporating the above strategies, I sleep seven to eight hours a night and wake up feeling refreshed most mornings. You can, too.

■ ■

Those are some mighty powerful stress-managing strategies. And in case you didn't notice, you can combine many of them simultaneously for synergistic effects. For instance, you could try deep abdominal breathing while going for a walk in nature, all while expressing what you're grateful for. Apply any number of these magical strategies and you'll be well on your way to supporting your adrenals and naturally elevating your energy levels for a life of vitality and happiness.

■ ■ ■ ■

Chapter Eleven

FINDING YOUR PASSION AND PURPOSE TO FEEL AWESOME

Why do you wake up in the morning? What on earth are you on this earth for?

Those are very lofty questions, and they might seem outside the scope of this book. However, I believe them to be two crucial questions you should ask yourself, especially if you're trying to revitalize your life.

If you've made it this far in the book, it means you've taken a hard look at all the happenings inside of your body that have been slowing you down and holding you back. Together, we've tackled your digestion and your blood alkalinity; we've also overhauled your exercise regimen and stress-management techniques. Hopefully, working on all of these things has you feeling better than ever before, but there's one last thing that will truly unlock the life you were meant to lead and the unstoppable energy that goes with it: inspiration.

As important as your physical well-being is, you need inspiration in your life. If not, what are you even getting in shape for? This might sound funny coming from a health expert, but it's just as important as any technical information I can share with you.

You have to have passion and purpose to really access the energy you're hoping for.

There never has been and never will be another "you" on this planet. You might not ever have thought about this, but there's a tremendous responsibility that comes with that, and thankfully, acknowledging this will make your life the amazing gift it is meant to be. In fact, this is a gift that you have to give back to the world. This isn't a chore, but a blessing. It can take you from shuffling through your days like a zombie to celebrating each and every breath you take. Your life is meant to be nothing less than awesome!

Accepting this is surely a bit daunting, and I know this from my own experience. Throughout my teenage years and into my early adulthood, I was certain that I was meant to be a professional soccer player. I was thoroughly devoted to the sport, even when my health really started to suffer and I felt horrid. It's simply who I was. However, when I turned 24, everything changed.

It's funny how it all crept up on me without me realizing it. By that point, I was becoming very critical of my performance on the field, so much so that I didn't even enjoy playing anymore. I was harder on myself than my coach was!

It was a shock to realize how truly unhappy I was. In fact, it was quite scary. I was meant to be a professional athlete; what else was there? With that question came a burning desire to find out what I really wanted for myself. By the time I eventually hung up my cleats, I understood that soccer simply wasn't enough for me; I wanted to serve the world, and the best way to do so was by sharing all that I learned in my mission to reclaim my health.

I took a long road to get there, but what I experienced was nothing less than an epiphany. I am so glad it hit me when it did, because I can't imagine what my life would be like if I'd continued playing soccer. I am so much happier with the life I live now, and every day I am inspired to keep going. My mission to help ten million people live healthier, fitter lives in the next few years keeps me going every day, and helps me to plow through any obstacles I am confronted with. After years and years of struggle and sorrow, I'm finally alive. Every day is an adventure that I'm eager to take on.

What will it take for you to experience this feeling?

Clearing Out the Clutter in Your Life

Your path to your passion doesn't have to be as dramatic as mine. I certainly don't want you to lose all of your hair or experience the existential crisis I had to go through. That was no fun. Then again, there's a good chance your life right now isn't much fun either.

Think about it: how much do you actually enjoy your life? It doesn't matter what you do for a living; you could be a CEO of a Fortune 500 company or a cashier at the gas station down the block. Is your daily life fulfilling? If so, that's fantastic. Just work on everything we've talked about so far and you'll be quite fine. However, if your daily life is a hassle, a bore, or even a nightmare, you have to do something about it. As long as this is the case, there's no way you can ever experience the energy that's been eluding you. It simply isn't possible.

It's so easy to lose ourselves in our sense of obligation, be it to our jobs or our families. We forget that in order to truly help or serve anyone else, we have to take care of ourselves first, and that means being dedicated to something we find truly fulfilling.

Some of you reading this may be stay-at-home parents. Although many might not see it this way, parenting your children is a full-time profession, especially if you've deliberately chosen not to have a full-time job. Over time, however, you can begin to exist solely for your children and find yourself exhausted. That's when you begin to buy books like this!

This isn't to say that you should go and find a job immediately. Rather, you owe it to yourself and your children to bring something into your life that really inspires you and restores the energy you expend taking care of your little ones. The same thing applies if you're working a 9-to-5 job behind a desk with a boss who runs you into the ground. It doesn't matter if you're a man or a woman, old or young. As long as you're giving all of yourself away to someone or something else, you will eventually feel drained and empty.

You might think this is easy for me to say. After all, I have a successful company and you're reading the book that I wrote. How can I possibly relate to your struggles?

What I'm sharing with you not only worked for me, but also for so many of my own clients. I want you to be just as fulfilled as we are.

Here's something you can do right now to start moving in that direction. Take out a piece of paper, and draw a line down the center. At the top of one column, write "5 percent," and at the top of the other, write "95 percent." In the 5 percent column, write down the things that you love to do or the things that you absolutely excel at. Anything goes, be it a sport, a hobby, or a profession. It could be something you've been amazing at in the past, or it could be something you're dying to try but never have.

In the 95 percent column, you're going to write all of the things that weigh you down. Feel free to really vent here, and write down every last thing that confounds, bothers, and depresses you. Don't hold anything back.

Chances are, you'll have quite a few more entries in this column than the other one, and that's a big indication that you're not living in alignment with your truest self. How could you possibly be happy if you have all of this stuff to contend with?

I have found this exercise to be so useful in every aspect of my life. When I applied it to my diet, I realized that I didn't like spending a lot of time in the kitchen, so I incorporated more smoothies into my daily schedule, as they were easy and quick to make without sacrificing taste or nutrition. With my business, I realized that I simply didn't enjoy micromanaging people, so I made it a point to empower every person I work with or employ with the tools they need to get the job done without consulting me any more than they absolutely need to.

The purpose of this exercise is to sift out the aspects of your life that you want to handle differently. It's hard to do that when you wake up in the morning with a million and one things to do. When you take a step back and do this exercise, you're able to manage your life with a little more perspective, especially when you apply a principle I call A.D.D.—*automate, delegate,* and *delete.*

Here's how it works: once you've finished your list, circle three items in your 95 percent column that you just don't want to deal with anymore. Now, look at them and start thinking about how you can automate, delegate, or delete them. For example, you might do all of your laundry at home, when the slight expense of having a laundromat take care of it for you might free up so much of your time. Automate that task by scheduling a weekly pickup. Similarly, hiring someone to clean your home or do your taxes can be a time saver as well. This is delegating. Finally, there will just be some items that you want to get rid of for good, avoidable hassles in your life that you just can't deal with anymore. Delete them.

It's a very simple exercise, but you'll be amazed by how much pressure and weight it takes off of your shoulders. Best of all, it gives you more time, space, and energy to devote to things in the 5 percent column. Just like that, you empower yourself to start living a more enjoyable life, one day at a time.

The Magic of Giving Back

There's a profound difference between my life as a soccer player and the one I live now as a fitness and nutrition coach. It all has to do with how much I contribute to the world around me.

If you flip back to the Introduction of this book, you'll see that I mention very early on how much helping people means to me. Whether you realize it or not, you probably feel the same way, too. It doesn't matter what your dream profession is. If you want to be an artist, part of your work involves delighting or provoking people in new and interesting ways with your art. If you want to be an engineer, you likely want to have a hand in building things that benefit society at large. Whatever you do, there's a strong possibility it involves some level of contribution.

In a 2013 paper out of the University of Exeter, 40 studies on volunteering from the 20 years prior found that helping others was associated with an increased sense of well-being, lower rates of depression, and a 22 percent reduction in the risk of dying early.[1]

It's very simple: helping others helps you become happier and healthier.

I didn't include this fact in the chapter on stress management, but it's something you should try next time you're feeling stressed-out. When the pressure hits and you're feeling low, instead of allowing yourself to slip into a funky mood, turn your thoughts to how you can help somebody else. I guarantee you it will heighten your mood every single time you do it. The best part of it is that someone will benefit from your generosity of spirit.

This shouldn't be surprising when you think about it, and hopefully, this factoid will make you think about the best way you can give back to your community and your world. I find the best way to do this is to practice what you love best and share it with everyone you can. Can you imagine what our world would be like if everyone was encouraged to do this? Just think about it.

When I was playing soccer, it was all about me, my achievements, and the money I was hoping to make. However, after a while I started to ask myself: *Am I making the world a better place?* I could tell you a million things that were wrong with the world around me, but I wasn't making an effort to change any of them. Now, I feel fulfilled and purposeful because I'm helping so many to live meaningful lives. My life now far eclipses anything I would have experienced if I had stayed on the field.

Mind you, you don't have to save the world, nor do you have to reach as many people as I do. Your calling might simply involve helping your family or the people in your surrounding community. What I'm certain of is that when you're doing what you're meant to be doing, you're not the only person who will benefit from it.

One Day at a Time

The last point I want to make is this: don't expect everything to change overnight. Unlocking your passion and changing your life is a process, and by giving yourself over to it every step of the way, you partake in a personal adventure that you'll never forget. It's in taking note of your progress that you build motivation that can last a lifetime and inspire others to do the same.

Everyone has a different story. Some people work a full career until their 50s and only then find their calling. Others abandon this 9-to-5 life early on and give their passion a go. Some inspired teenagers dedicate their lives to their passion before they even hit college, and they become success stories before they even land their first job.

The key here is to be patient with yourself, and wake up every day dedicated to making it better than the one before.

Progress is more important than you might realize. Shortly before this book was published, I had to drive to a meeting in the heart of downtown Toronto, about 45 minutes away from my home. I was hoping to go and take care of my business and be back at my home office before lunch, but it wasn't meant to be. There was an accident on the highway that day, and my commute ended up lasting a whopping two and half hours. As you can imagine, I wasn't thrilled.

This little anecdote might reflect how you feel every day. That feeling of stagnancy can wear you down, so it's important to really pay attention to the work you're doing and where it's taking you. In the last chapter, I mentioned keeping a gratitude journal, and you might want to consider keeping a general journal that makes note of your accomplishments as you work toward your health, stress-management, and professional goals. As you keep track of your progress, you'll be filled with a steadily growing sense of accomplishment that will fuel you as you go.

In *You Can Heal Your Life,* the renowned book that started the very company that published this book you now hold, Louise Hay talks about how fatigue can be born of a lack of love for what you do and a missing zest for life. You can find it, but it may not happen as quickly as you're hoping. That's why it's crucial that you're patient and gentle with yourself. That's when the miracles happen.

■ ■

We've covered so much ground together, and it's my hope that you're looking at your life with new eyes now. Your life is a treasure, and for so long, you haven't treated it as such. You've been hard on yourself, barely getting the sleep you need and working yourself far beyond your limits. To cope, you've eaten the worst foods possible and pumped yourself full of toxic "pick-me-ups" that slowly kill you. In the process, you've become a shell of yourself, forever wondering why life has to be so difficult.

I promise you that the approach in these pages will erase all of this and give you a new lease on life. If you feel better after just seven days of following these methods, can you imagine how you'll feel after a month? How about a year?

I hope you've come to realize that your happiness won't come about from getting a ripped body or eating the healthiest diet imaginable. Neither will it come from a bounty of money and supposed success. It's all about finding balance in yourself, and allowing your natural energy to power you through your days. There's a powerhouse in you just waiting to light up the world. You're closer than you think.

■ ■ ■ ■

ENDNOTES

Chapter 1

1. Heckman, M. A., K. Sherry, D. Mejia, and E. Gonzalez. "Energy drinks: An assessment of their market size, consumer demographics, ingredient profile, functionality, and regulations in the United States." *Comprehensive Reviews in Food Science and Food Safety* 9, no. 3 (2010): 303–317.

2. Rohleder, Nicolas, Ljiljana Joksimovic, Jutta M. Wolf, and Clemens Kirschbaum. "Hypocortisolism and increased glucocorticoid sensitivity of pro-inflammatory cytokine production in Bosnian war refugees with posttraumatic stress disorder."*Biological Psychiatry* 55, no. 7 (2004): 745–751.

3. McEwen, Bruce S., and Teresa Seeman. "Protective and damaging effects of mediators of stress: elaborating and testing the concepts of allostasis and allostatic load." *Annals of the New York Academy of Sciences* 896, no. 1 (1999): 30–47.

4. Elenkov, Ilia J., and George P. Chrousos. "Stress hormones, Th1/Th2 patterns, pro/anti-inflammatory cytokines and susceptibility to disease." *Trends in Endocrinology & Metabolism* 10, no. 9 (1999): 359–368.

Chapter 2

1. Eaton, S. Boyd. "The ancestral human diet: What was it and should it be a paradigm for contemporary nutrition?" *Proceedings of the Nutrition Society* 65, no. 1 (2006): 1–6.

2. Slavin, J. L. "Position of the American Dietetic Association: Health implications of dietary fiber." *Journal of the American Dietetic Association* 108, no. 10 (2008): 1716–1731.

3. Johnson, Rachel K., Lawrence J. Appel, Michael Brands, Barbara V. Howard, Michael Lefevre, Robert H. Lustig, Frank Sacks, Lyn M. Steffen, and Judith Wylie-Rosett. "Dietary sugars intake and cardiovascular health: A scientific statement from the American Heart Association." *Circulation* 120, no. 11 (2009): 1011–1020.

4. Clauson, Kevin A., Kelly M. Shields, Cydney E. McQueen, and Nikki Persad. "Safety issues associated with commercially available energy drinks." *Pharmacy Today* 14, no. 5 (2008): 52–64.

Chapter 3

1. Cseuz, Regina Maria, Istvan Barna, Tamas Bender, and Jürgen Vormann. "Alkaline mineral supplementation decreases pain in rheumatoid arthritis patients: A pilot study." *Open Nutrition Journal* 2 (2008): 100–105.

2. Goraya, Nimrit, and Donald E. Wesson. "Does correction of metabolic acidosis slow chronic kidney disease progression?" *Current Opinion in Nephrology and Hypertension* 22, no. 2 (2013): 193–197.

3. Adeva, María M., and Gema Souto. "Diet-induced metabolic acidosis." *Clinical Nutrition* 30, no. 4 (2011): 416–421.

4. Koufman, Jamie A., and Nikki Johnston. "Potential benefits of pH 8.8 alkaline drinking water as an adjunct in the treatment of reflux disease." *Annals of Otology Rhinology and Laryngology-Including Supplements* 121, no. 7 (2012): 431.

5. Kanbara, Aya, Yoshisuke Miura, Hideyuki Hyogo, Kazuaki Chayama, and Issei Seyama. "Effect of urine pH changed by dietary intervention on uric acid clearance mechanism of pH-dependent excretion of urinary uric acid." *Nutrition Journal* 11, no. 1 (2012): 39.

6. Tucker, Katherine L., Marian T. Hannan, and Douglas P. Kiel. "The acid-base hypothesis: Diet and bone in the Framingham Osteoporosis Study." *European Journal of Nutrition* 40, no. 5 (2001): 231–237.

Chapter 4

1. Kaats, Gilbert R., Dennis Pullin, and Larry K. Parker. "The short term efficacy of the ALCAT test of food sensitivities to facilitate changes in body composition and self-reported disease symptoms: A randomized controlled study." *Health & Medicine Research Foundation, Baylor Sports Medicine Institute (Houston). In Publication* (1996).

2. Akmal, Mohammed, Saeed Ahmed Khan, and Abdul Qayyum Khan. "The Effect of the ALCAT Test diet therapy for food sensitivity in patients with obesity."*Middle East Journal of Family Medicine* 7, no. 3 (2009).

3. Pitkin, R. M., L. H. Allen, L. B. Bailey, and M. Bernfield. "Dietary reference intakes for thiamin, riboflavin, niacin, vitamin B_6, folate, vitamin B_{12}, pantothenic acid, biotin and choline." (2000): 196–305.

Chapter 5

1. Dugdale, David C., III. "What causes bone loss?" *MedlinePlus*. U.S. National Library of Medicine. National Institutes of Health (2012).

2. Muehlhoff, E., A. Bennett, and D. McMahon. *Milk and Dairy Products in Human Nutrition*. Food and Agriculture Organization of the United Nations: Rome, 2013.

3. Paspati, I., A. Galanos, and G. P. Lyritis. "Hip fracture epidemiology in Greece during 1977–1992." *Calcified Tissue International* 62, no. 6 (1998): 542–547.

4. Muehlhoff, E., A. Bennett, and D. McMahon. *Milk and Dairy Products in Human Nutrition*. Food and Agriculture Organization of the United Nations: Rome, 2013.

5. Lau, E. M., and C. Cooper. "Epidemiology and prevention of osteoporosis in urbanized Asian populations." *Osteoporosis International* 3, no. 1 (1993): 23–26.

6. Ho, Suzanne C., Edith Lau, Jean Woo, Aprille Sham, Kai Ming Chan, Simon Lee, and Ping Chung Leung. "The prevalence of osteoporosis in the Hong Kong Chinese female population." *Maturitas* 32, no. 3 (1999): 171–178.

7. Lau, E. M. "Admission rates for hip fracture in Australia in the last decade: The New South Wales scene in a world perspective." *The Medical Journal of Australia* 158, no. 9 (1993): 604–606.

8. Fujita, Takuo, and Masaaki Fukase. "Comparison of osteoporosis and calcium intake between Japan and the United States." *Experimental Biology and Medicine* 200, no. 2 (1992): 149–152.

9. Bauer, Richard L. "Ethnic differences in hip fracture: A reduced incidence in Mexican Americans." *American Journal of Epidemiology* 127, no. 1 (1988): 145–149.

10. Bohannon, Arline. "Osteoporosis and African American women." *Journal of Women's Health & Gender-Based Medicine* 8, no. 5 (1999): 609–615.

11. Howell, Edward. *Enzyme Nutrition: The Food Enzymes Concept*. Penguin.com (1985).

12. Saunders, C. W. "The nutritional value of chlorophyll as related to hemoglobin formation." In *Proceedings of the Society for Experimental Biology and Medicine. Society for Experimental Biology and Medicine (New York, NY)* 23, no. 8 (1926): 788–789.

13. Smith, Lawrence W. "Chlorophyll: An experimental study of its water-soluble derivatives: I. Remarks upon the history, chemistry, toxicity and antibacterial properties of water-soluble chlorophyll derivatives as therapeutic agents." *The American Journal of the Medical Sciences* 207, no. 5 (1944): 647–654.

14. Patek, Jr., Arthur J. "Chlorophyll and regeneration of the blood: Effect of administration of chlorophyll derivatives to patients with chronic hypochromic anemia." *Archives of Internal Medicine* 57, no. 1 (1936): 73.

15. Hung, Hsin-Chia, Kaumudi J. Joshipura, Rui Jiang, Frank B. Hu, David Hunter, Stephanie A. Smith-Warner, Graham A. Colditz, Bernard Rosner, Donna Spiegelman, and Walter C. Willett. "Fruit and vegetable intake and risk of major chronic disease." *Journal of the National Cancer Institute* 96, no. 21 (2004): 1577–1584.

16. Steinmetz, Kristi A., and John D. Potter. "Vegetables, fruit, and cancer prevention: A review." *Journal of the American Dietetic Association* 96, no. 10 (1996): 1027–1039.

17. Young, Vernon R., and Peter L. Pellett. "Plant proteins in relation to human protein and amino acid nutrition." *The American Journal of Clinical Nutrition* 59, no. 5 (1994): S1203–S1212.

18. Kannel, William B., Daniel McGee, and Tavia Gordon. "A general cardiovascular risk profile: The Framingham Study." *The American Journal of Cardiology* 38, no. 1 (1976): 46–51.

19. Holmen, Jostein, Kristian Midthjell, Øystein Krüger, Arnulf Langhammer, Turid Lingaas Holmen, Grete H. Bratberg, Lars Vatten, and Per G. Lund-Larsen. "The Nord-Trøndelag Health Study 1995–97 (HUNT 2): Objectives, contents, methods and participation." *Norsk epidemiologi* 13, no. 1 (2003): 19–32.

20. Huttenlocher, P. R., A. J. Wilbourn, and J. M. Signore. "Medium-chain triglycerides as a therapy for intractable childhood epilepsy." *Neurology* 21, no. 11 (1971): 1097.

21. Mensink, Ronald P., Peter L. Zock, Arnold D. M. Kester, and Martijn B. Katan. "Effects of dietary fatty acids and carbohydrates on the ratio of serum total to HDL cholesterol and on serum lipids and apolipoproteins: A meta-analysis of 60 controlled trials." *The American Journal of Clinical Nutrition* 77, no. 5 (2003): 1146–1155.

22. Bogani, Paola, Claudio Galli, Marco Villa, and Francesco Visioli. "Postprandial anti-inflammatory and antioxidant effects of extra virgin olive oil." *Atherosclerosis* 190, no. 1 (2007): 181–186.

23. "You Can Control Your Cholesterol: A Guide to Low-Cholesterol Living." Merck & Co. Inc.

24. Visioli, Francesco, and Tory M. Hagen. "Nutritional strategies for healthy cardiovascular aging: Focus on micronutrients." *Pharmacological Research* 55, no. 3 (2007): 199–206.

25. Ho, Emily, Thomas W-M. Boileau, and Tammy M. Bray. "Dietary influences on endocrine–inflammatory interactions in prostate cancer development." *Archives of Biochemistry and Biophysics* 428, no. 1 (2004): 109–117.

26. Brenna, J. Thomas, Norman Salem, Jr., Andrew J. Sinclair, and Stephen C. Cunnane. "ISSFAL Official Statement Number 5 α–Linolenic Acid Supplementation and Conversion to n-3 Long Chain Polyunsaturated Fatty Acids in Humans." International Society for the Study of Fatty Acids and Lipids (2009).

27. Sanders, Thomas A. B. "DHA status of vegetarians." *Prostaglandins, Leukotrienes, and Essential Fatty Acids* 81, no. 2: 137–141.

28. Lukiw, Walter J., Jian-Guo Cui, Victor L. Marcheselli, Merete Bodker, Anja Botkjaer, Katherine Gotlinger, Charles N. Serhan, and Nicolas G. Bazan. "A role for docosahexaenoic acid–derived neuroprotectin D1 in neural cell survival and Alzheimer disease." *Journal of Clinical Investigation* 115, no. 10 (2005): 2774–2783.

29. Farzaneh-Far, Ramin, Jue Lin, Elissa S. Epel, William S. Harris, Elizabeth H. Blackburn, and Mary A. Whooley. "Association of marine omega-3 fatty acid levels with telomeric aging in patients with coronary heart disease." *JAMA* 303, no. 3 (2010): 250–257.

Chapter 7

1. Kunze, R., K. Ransberger, et al. "Humoral immunomodulatory capacity of proteases in immune complex decomposition and formation." First International Symposium on Combination Therapies, Washington, D.C., 1991.

2. Kumar, Ratan, S. K. Grover, H. M. Divekar, A. K. Gupta, Radhey Shyam, and K. K. Srivastava. "Enhanced thermogenesis in rats by Panax ginseng, multivitamins and minerals." *International Journal of Biometeorology* 39, no. 4 (1996): 187–191.

3. Kelly, Gregory S. "Nutritional and botanical interventions to assist with the adaptation to stress." *Alternative Medicine Review: A Journal of Clinical Therapeutics* 4, no. 4 (1999): 249–265.

4. Chacón de Popovici, Gloria. *MACA (Lepidium peruvianum Chacón): Millenarian Peruvian Food Plant, With Highly Nutritional and Medicinal Properties.* Lima, Peru: Gráfica Mundo. (2001): 1–337.

5. Gonzales, Gustavo F., Carla Gonzales, and Cynthia Gonzales-Castaneda. "Lepidium meyenii (Maca): a plant from the highlands of Peru–from tradition to science." *Forschende Komplementärmedizin/Research in Complementary Medicine* 16, no. 6 (2009): 373–380.

6. Vecera, Rostislav, Jan Orolin, Nina Skottova, Ludmila Kazdova, Olena Oliyarnik, Jitka Ulrichova, and Vilim Simanek. "The influence of maca (Lepidium meyenii) on antioxidant status, lipid and glucose metabolism in rat." *Plant Foods for Human Nutrition* 62, no. 2 (2007): 59–63.

7. Panossian, A., G. Wikman, and J. Sarris. "Rosenroot (*Rhodiola rosea*): Traditional use, chemical composition, pharmacology and clinical efficacy." *Phytomedicine* 17, no. 7 (2010): 481–493.

8. De Bock, Katrien, Bert O. Eijnde, Monique Ramaekers, and Peter Hespel. "Acute Rhodiola rosea intake can improve endurance exercise performance." *International Journal of Sport Nutrition and Exercise Metabolism* 14, no. 3 (2004): 298–307.

9. Zhang, Zhang-jin, Yao Tong, Jun Zou, Pei-jie Chen, and Ding-hai Yu. "Dietary supplement with a combination of Rhodiola crenulata and Ginkgo biloba enhances the endurance performance in healthy volunteers." *Chinese Journal of Integrative Medicine* 15 (2009): 177–183.

10. Samy, Ramar Perumal, Peter Natesan Pushparaj, and Ponnampalam Gopalakrishnakone. "A compilation of bioactive compounds from Ayurveda." *Bioinformation* 3, no. 3 (2008): 100.

11. Shenoy, Shweta, Udesh Chaskar, Jaspal S. Sandhu, and Madan Mohan Paadhi. "Effects of eight-week supplementation of Ashwagandha on cardiorespiratory endurance in elite Indian cyclists." *Journal of Ayurveda and Integrative Medicine* 3, no. 4 (2012): 209.

Chapter 10

1. Geer, James H., Gerald C. Davison, and Robert I. Gatchel. "Reduction of stress in humans through nonveridical perceived control of aversive stimulation." *Journal of Personality and Social Psychology* 16, no. 4 (1970): 731.

2. Bollini, Annie M., Elaine F. Walker, Stephan Hamann, and Lisa Kestler. "The influence of perceived control and locus of control on the cortisol and subjective responses to stress." *Biological Psychology* 67, no. 3 (2004): 245–260.

3. Baker, Susan P., Joanne E. Brady, Dennis F. Shanahan, and Guohua Li. "Aviation-related injury morbidity and mortality: Data from US health information systems." *Aviation, Space, and Environmental Medicine* 80, no. 12 (2009): 1001–1005.

4. Bushman, Brad J., Roy F. Baumeister, and Angela D. Stack. "Catharsis, aggression, and persuasive influence: Self-fulfilling or self-defeating prophecies?" *Journal of Personality and Social Psychology* 76, no. 3 (1999): 367–376.

5. Hartig, Terry, Marlis Mang, and Gary W. Evans. "Restorative effects of natural environment experiences." *Environment and Behavior* 23, no. 1 (1991): 3–26.

6. Pretty, Jules, Jo Peacock, Rachel Hine, M. Sellens, N. South, and M. Griffin. "Green exercise in the UK countryside: Effects on health and psychological

well-being, and implications for policy and planning." *Journal of Environmental Planning and Management* 50, no. 2 (2007): 211–231.

7. Mainster, Martin A. "Violet and blue light blocking intraocular lenses: Photoprotection versus photoreception." *British Journal of Ophthalmology* 90, no. 6 (2006): 784–792.

Chapter 11

1. Jenkinson, Caroline E., Andy P. Dickens, Kerry Jones, Jo Thompson-Coon, Rod S. Taylor, Morwenna Rogers, Clare L. Bambra, Iain Lang, and Suzanne H. Richards. "Is volunteering a public health intervention? A systematic review and meta-analysis of the health and survival of volunteers." *BMC Public Health* 13, no. 1 (2013): 1–10.

■ ■ ■ ■

FREE GIFT FOR YOU

I'd like to thank you for picking up this book and demonstrating that you are someone who is committed to living a full, energized, and healthy life. To help you meet these goals and to further support you in your journey, I've prepared the following 3 free gifts for you:

- **Food Labeling 101**—chances are, you will still have some packaged foods from time to time. This bonus shows you how to understand those tricky food labels so you don't get taken for a ride.
- **8 Energy-Boosting Desserts**—these yummy and healthy desserts will finally allow you to have your cake and eat it, too, without the dangers of sugar, dairy, or gluten.
- **Yoga for Energy**—a 15-minute yoga routine to relax and rejuvenate your body.

To download your 3 FREE All-Day Energy Diet bonuses, simply go to:

www.alldayenergydiet.com/3gifts

RECOMMENDED RESOURCES

Throughout this book, I've reiterated the importance of alkalinity for improving your health and taking your energy to new levels. I've also mentioned a number of very helpful supplements and tactics that boost your energy, improve your health, and ease your stress. On this page, you can find a list of my top recommendations for you to get your hands on. I've also included some additional books and resources that you might find helpful.

Energy Greens—www.alldayenergydiet.com/20greens

The tastiest supergreens powder you'll ever taste. Enjoy 8 powerful superfoods in less than 30 seconds a day in the most delicious and convenient way possible. And as a valued reader, I'm hooking you up with a 20% discount. Just go to the website above.

Yoga for Athletes—www.alldayenergydiet.com/yoga

Enjoy the body rejuvenating, stress-busting 25-minute yoga routine I created for the men's soccer program at the University of Toronto.

Yuri-approved digestive enzymes—www.alldaydenergydiet.com/enzymes

Yuri-approved probiotics—www.alldayenergydiet.com/probiotics

Exercise Library—www.alldayenergydiet.com/exercises

In the exercise, I alluded to a number of exercises and workouts. You'll be happy to know that I've recorded instructional videos for each one of the exercises so that you know exactly how to do them. Just go to the URL mentioned above.

All-Day Energy Smoothies and Juices—www.alldayenergy diet.com/smoothies

Get 54 delicious smoothie and juicing recipes that have been tried, tested, and approved by my 3 young kids. If you're looking for a simple way to get excellent nutrition, then you'll love these liquid recipes.

■ ■ ■ ■

ABOUT THE AUTHOR

Yuri Elkaim is a registered holistic nutritionist and fitness expert most famous for helping people quickly enjoy all-day energy and amazing health without gimmicks or fad diets. He's also well known for getting people into amazing athletic shape no matter where they started and without killing themselves in the gym. His uplifting message of health has already transformed the lives of 500,000 people worldwide, and he's on a mission to help at least ten million more.

Yuri holds a high honours degree in physical education and health/kinesiology from the University of Toronto, was a two-time All-Canadian soccer standout and former professional soccer player, and for seven seasons acted as the head strength-and-conditioning and nutrition coach for the men's soccer programme at the University of Toronto.

Yuri is the professor of Super Nutrition Academy; the author of *Eating for Energy* and *The Total Wellness Cleanse;* and the creator of more than 130 workout programmes, including Fitter U, Fitter U Fitness, Amazing Abs Solution and Treadmill Trainer.

www.alldayenergydiet.com

NOTES

We hope you enjoyed this Hay House book. If you'd like to receive our online catalog featuring additional information on Hay House books and products, or if you'd like to find out more about the Hay Foundation, please contact:

Hay House UK, Ltd., Astley House, 33 Notting Hill Gate, London W11 3JQ
Phone: 0-20-3675-2450 • *Fax:* 0-20-3675-2451
www.hayhouse.co.uk• www.hayfoundation.org

III

Published and distributed in the United States by: Hay House, Inc., P.O. Box 5100, Carlsbad, CA 92018-5100
Phone: (760) 431-7695 or (800) 654-5126
Fax: (760) 431-6948 or (800) 650-5115
www.hayhouse.com®

Published and distributed in Australia by: Hay House Australia Pty. Ltd., 18/36 Ralph St., Alexandria NSW 2015 • *Phone:* 612-9669-4299
Fax: 612-9669-4144 • www.hayhouse.com.au

Published and distributed in the Republic of South Africa by:
Hay House SA (Pty), Ltd., P.O. Box 990, Witkoppen 2068
Phone/Fax: 27-11-467-8904 • www.hayhouse.co.za

Published in India by: Hay House Publishers India, Muskaan Complex, Plot No. 3, B-2, Vasant Kunj, New Delhi 110 070 • *Phone:* 91-11-4176-1620
Fax: 91-11-4176-1630 • www.hayhouse.co.in

Distributed in Canada by: Raincoast Books, 2440 Viking Way, Richmond, B.C. V6V 1N2 • *Phone:* 1-800-663-5714 • *Fax:* 1-800-565-3770 • www.raincoast.com

III